Junji Tobias Märkl

Investigation of Magnetic Adatoms with Scanning Tunneling Techniques

Experimental Condensed Matter Physics
Band 19

Herausgeber
Physikalisches Institut

Prof. Dr. Hilbert von Löhneysen
Prof. Dr. Alexey Ustinov
Prof. Dr. Georg Weiß
Prof. Dr. Wulf Wulfhekel

Eine Übersicht aller bisher in dieser Schriftenreihe erschienenen Bände finden Sie am Ende des Buchs.

Investigation of Magnetic Adatoms with Scanning Tunneling Techniques

by
Junji Tobias Märkl

KIT Scientific Publishing

Dissertation, Karlsruher Institut für Technologie (KIT)
Fakultät für Physik
Tag der mündlichen Prüfung: 3. Juli 2015
Referenten: Prof. Dr. Wulf Wulfhekel, Prof. Dr. Georg Weiß

Impressum

KIT Scientific Publishing

Karlsruher Institut für Technologie (KIT)
KIT Scientific Publishing
Straße am Forum 2
D-76131 Karlsruhe

KIT Scientific Publishing is a registered trademark of Karlsruhe Institute of Technology. Reprint using the book cover is not allowed.

www.ksp.kit.edu

Print on Demand 2015

ISSN 2191-9925
ISBN 978-3-7315-0435-1
DOI 10.5445/KSP/1000049666

Contents

Chapter 1.

1 Introduction

For decades, magnetic recording has been the dominant technology for information storage on hard disk drives. The advantage of high storage density for a low price has long outweighed the comparably low access speeds and data transfer rates. Although solid state disks have become more and more popular in recent years, magnetic hard disk technology is continuously improved by ongoing research. New methods to increase the information density, such as *shingled magnetic recording* [1] or *heat assisted magnetic recording* [2], are being developed to circumvent physical limitations of current generation disks, partially even sacrificing speed for storage density. Simultaneously, radically different concepts for magnetic data storage, such as *racetrack memory* [3] were proposed, studied and developed.

However, although further technological progress can be expected, in all of these technologies the size of one magnetic bit will eventually approach scales where the superparamagnetic limit cannot be circumvented anymore. This limit designates the size of a grain of magnetic material, below which thermal fluctuations lead to a rapid loss of the magnetization orientation. Even before reaching this limit, a high density of magnetic bits poses new difficulties for writing information to a bit without affecting its neighbors. The crucial task during this endeavor is to find magnetic systems with a suitable **magnetic anisotropy** so that their magnetization — the degree of freedom that encodes the stored information — shows a stability sufficient for applications. Should research in this direction still be successful, the quest for even further miniaturization will find its ultimate limit at the atomic scale, where quantum mechanics determines the behavior of magnetic objects.

Instead of studying progressively smaller and smaller clusters of magnetic material, one can choose a different approach: direct investigation of actual atomic-scale systems such as single magnetic atoms or molecules. While there are several techniques to study large ensembles of atomic-sized objects either in bulk or on surfaces, only few are capable of addressing them individually. Among those, **scanning tunneling microscopy** (STM) is certainly the most versatile as it allows to spatially characterize and manipulate the sample surface on sub-nanometer length scales. Furthermore, it offers a variety of spectroscopy techniques to study excitations, electronic and even magnetic properties.

Research on single magnetic atoms on surfaces with STM started in 1998, when cobalt atoms on the (111) surface of gold were studied [4]. After inelastic scanning tunneling spectroscopy (ISTS) was established in the same year by revealing molecular vibration modes [5], the spin degree of freedom of magnetic atoms was investigated for the first time in 2004 [6]. Since then, numerous studies – both experimentally and theoretically – have been carried out in order to quantify the anisotropy of these systems, as well as identify and understand the processes that govern the spin dynamics. Although the anisotropy barriers of single adatoms were often found to be much larger than for bulk materials (e.g., [7]), in almost all cases it was found that the lifetime of the magnetic ground state(s) is either extremely short, or the formation of a Kondo-singlet was observed. Both effects prevent using an atom's spin as a magnetic bit. In contrast, small antiferromagnetic structures of just a few atoms could be built by atomic manipulation that showed long-term stability of their spin orientation over many hours [8].

Among the single atoms, however, one system stood out from all other investigations: In holmium atoms on a platinum-(111) surface, ground state lifetimes of up to several hundred seconds were observed [9], which is many orders of magnitude longer than in other single adatoms. This experimental finding was the result of sustained research started by Schuh et al. [10] which was continued in the present thesis. As will be laid out in detail, the long lifetimes are a consequence of a fortunate combination of atomic properties and symmetries of the substrate.

In this thesis, first the experimental technique (STM), its various operational modes, as well as the experimental setup will be introduced in chapter 2. Additionally, a section of that chapter describes the design, construction and testing of a superconducting coil which was manufactured as part of this thesis for the purpose of providing high magnetic fields in one of the STM setups. Chapter 3 focuses on a theoretical description for adatoms, which provides a powerful framework to predict the overall character of the magnetic properties. The central topic of Ho/Pt(111) will be presented in chapter 4. There, a summary of the previous findings of T. Schuh will be given and the subsequent experimental and theoretical progress will be discussed in detail. Two complementary projects to study single magnetic atoms (cobalt and cerium) on an insulating copper nitride layer are portrayed in chapter 5. Lastly, concluding remarks will be given in chapter 6.

Chapter 2.
Experimental Methods

Scanning tunneling microscopy (STM)[1] *has undoubtedly become an invaluable experimental tool for condensed matter physics. Since its invention by G. Binnig and H. Rohrer in 1982 [11] it has seen many improvements and extensions that enable it to measure not only topographic features but also the local electronic density of states (LDOS) and excitation spectra. Furthermore, the invention of spin-polarized STM allows to retrieve local magnetic information using the tunneling magnetoresistance effect (TMR).*

The first section of this chapter describes the fundamentals of these techniques and some aspects of the technical implementation; the second part of the chapter will present the STM setup (the "JT-STM") that was used for the experiments. That section will conclude with the description of the design, construction, and testing of a superconducting magnet. It was built during this thesis for another new STM facility that will be used in the future for some of the samples studied in this thesis.

2.1. Scanning Tunneling Microscopy

Although STM was invented in the early 1980s [11], tunnel junctions have been already studied since roughly 1960. A theoretical understanding of the peculiarities of tunnel junctions and their spectroscopic capabilities was therefore developed long before the invention of STM. Despite the different electrode geometry in STM, many of the concepts from planar tunnel junctions can still be applied.

In all these tunneling experiments the quantity being measured is the tunneling current and its dependency on various external parameters, such as bias voltage, external

[1] Throughout this thesis the abbreviation "STM" will be used for the technique (microscopy), as well as for the instrument (microscope). The distinction should always be obvious.

fields, temperature, or — specifically in the case of STM — position of the tip electrode. From this current measurement, many sample properties can be derived, as will be briefly described in the following sections. More detailed treatments of electron tunneling in general, and STM in particular, can be found in textbooks devoted to these topics, e.g., [12–14].

2.1.1. Fundamentals

The discovery of quantum tunneling is almost as old as quantum mechanics itself, and was very important for the explanation of classically impossible phenomena, such as the α-decay. It has therefore been included in virtually any textbook on basic quantum mechanics, and shall only be described here in the most limited form.

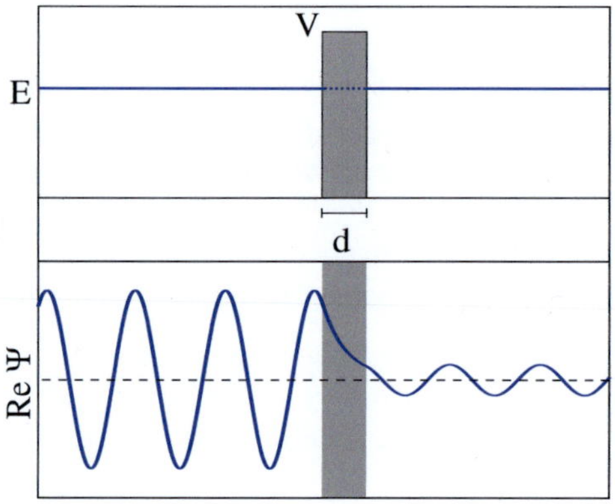

Figure 2.1.: *Illustration of the tunneling effect of a free particle impinging onto a potential barrier from the left. The particle's energy E, the barrier height V, and its width d influence the transmission coefficient which determines the amplitude of the transmitted wave function.*

For planar tunnel junctions, the mathematical description is usually reduced to a one-dimensional treatment that can be solved analytically. In the simplest case, a (quasi-) free particle with energy E impinges onto a rectangular potential barrier that separates two regions as presented in fig. 2.1. The particle then has a nonzero probability to tunnel through the barrier and continue its path on the other side. The associated tunnel probability τ depends on the particle's energy, as well as the barrier height V and width d:

$$\tau = \left(1 + \frac{V^2 \sinh^2(\kappa d)}{4E(V-E)}\right)^{-1} \tag{2.1a}$$

$$\approx 16\varepsilon(1-\varepsilon)\exp(-2\kappa d), \qquad \text{for } \kappa d \gg 1 \text{ and } \varepsilon = E/V \tag{2.1b}$$

where $\kappa = \sqrt{2m(V-E)}/\hbar$. Two qualitative aspects of a tunnel junction can easily be understood from the approximate formula: The tunnel probability depends exponentially on the distance d between the electrodes (= width of the barrier). In addition, there is an energy dependence which will favor the tunneling of higher energy particles.

The transition from this simple problem to the actual situation of a tunnel junction (both planar and STM) is usually nontrivial: Now, the tunneling particles are electrons that form a band structure in the electrodes. These, in turn, are in most cases made up of different materials, i.e., they have different work functions that determine the height and shape of the tunneling barrier. Furthermore, nonzero temperature results in Fermi-Dirac-like occupation of the available states in both electrodes. Last but not least, there is a nonzero bias voltage that creates an electric field across the barrier and leads to preferred tunneling in one direction.

Still, an argument by Bardeen can be used [15] to hide most of the complexity of the tunneling process in the matrix element M_{fi} of a current operator that represents tunneling between initial (i) and final (f) state. This gives a simple expression resembling Fermi's golden rule and describes the elastic tunneling current I when a bias voltage V is applied (see also fig. 2.2):

$$\begin{aligned} I =& \frac{2\pi e}{\hbar} \sum_{i,f} [f(E_{\mathrm{i}})(1 - f(E_{\mathrm{f}} + eV)) \\ & - f(E_{\mathrm{f}} + eV)(1 - f(E_{\mathrm{i}}))] \, |M_{\mathrm{fi}}(V)|^2 \delta(E_{\mathrm{i}} - E_{\mathrm{f}}) \\ =& \frac{2\pi e}{\hbar} \sum_{i,f} (f(E_{\mathrm{i}}) - f(E_{\mathrm{f}} + eV)) |M_{\mathrm{fi}}(V)|^2 \delta(E_{\mathrm{i}} - E_{\mathrm{f}}), \end{aligned} \tag{2.2}$$

where $f(E)$ denotes the Fermi function, and all energies are measured from the Fermi level of the respective electrode. Alternatively, the tunneling current can be expressed by using the electronic density of states $\rho_{1,2}$ in each electrode. This formulation implicitly assumes that M only depends on energy, i.e., other possible dependencies on wave vector and spin are neglected. Then, eq. (2.2) can be rewritten as:

$$I(V) = \frac{4\pi e}{\hbar} \int_{-\infty}^{\infty} \rho_1(E)\rho_2(E + eV)(f(E) - f(E + eV))|M(E,V)|^2 dE. \tag{2.3}$$

A similar result was obtained by Tersoff and Hamann [16] who accounted for the shape of an STM tip by assuming spherically symmetric orbitals as its electronic wave functions. They then found the tunneling current to be proportional to the sample density of states at the position of the tip. The above equation turns out to be useful for the explanations of the spectroscopy modes (section 2.1.2).

In contrast to planar tunnel junctions, where a thin insulating layer forms the barrier, the STM geometry directly employs vacuum to separate the two electrodes — tip (T) and sample (S). The technical implementation (see section 2.1.4) then allows to vary not only the lateral position of the tip for imaging and mapping purposes, but also the distance, thereby extending the possibilities by enabling measurements with variable conductance. In the following, the description will only focus on STM and its characteristics, especially keeping in mind the consequences that arise from being able to freely position its tip electrode.

As mentioned above, all the sample's physical properties of interest will be derived from the measured current. During the actual scanning of the surface, the so-called "constant current mode" is commonly used in which a feedback loop adjusts the tip-sample distance z. This variable distance can then be interpreted as the topographical height of the sample. However, on the basis of eq. (2.3) it is more accurate to state that it reflects the local density of states of the sample near the Fermi level. It is crucial to keep this in mind when comparing "heights" of objects on the surface. For example, adsorbates with a low density of states can appear as depressions in the topography although they certainly protrude from the surface. Furthermore, at different bias voltage, this contrast may even be reversed. Thus, a meaningful height can only be assigned to surface structures made from the same material, like reconstructions, or steps. Any other object will have an *apparent* height that may depend on the essential tunneling conditions. Hence, this information should always be stated for comparisons.

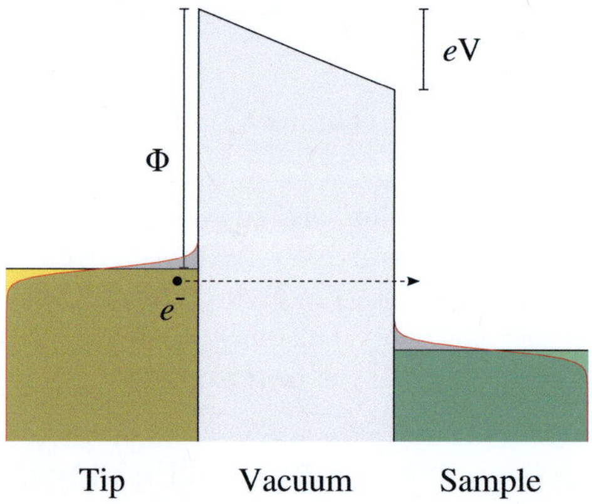

Figure 2.2.: *Schematic view of the simplified tunnel junction: A bias voltage V causes a tunnel current to flow; in this case electrons tunnel from the tip into the sample. The red curve on both sides represents the Fermi function, with the shaded area indicating the occupied states in each electrode.*

2.1.2. Spectroscopy Modes

Scanning Tunneling Spectroscopy (STS)

Equation 2.3 shows the tunneling current's dependence on the integrated local electronic density of states (LDOS) of the sample ρ_S. While the tip DOS ρ_T also influences the current, it is often approximated to be constant close to the Fermi level. Together with the assumption of an only weakly energy-dependent tunnel matrix element M, i.e., approximately constant in the energy interval of interest, one can derive from eq. (2.3)

$$\sigma(V_0) := \left.\frac{dI}{dV}\right|_{V_0} \propto \rho_T \; |M|^2 \; \rho_S(E_F + eV_0). \tag{2.4}$$

This equation implies that — given that certain assumptions hold — a measurement of the differential conductance σ directly allows to probe the LDOS. In such a measurement the I(V) curve at a given tip position will be recorded from which one can derive the differential conductance. This σ(V) curve then yields information about the energy at which the LDOS shows interesting features as well as their relative intensities. For higher voltages comparable to the work function, the energy dependent tunneling probability leads to a roughly exponentially growing current, a fact that must be taken into account for the interpretation of dI/dV curves.

Even more interesting possibilities arise when combining this point spectroscopy mode (commonly called scanning tunneling spectroscopy, STS) with the imaging capabilities. Scanning at a voltage that was found to show certain electronic features allows to detect spatial variations of the differential conductance which originate from locally different electronic properties. Among those, there can be, e.g., surface states, chemical contrast, different crystal structure, different adsorption sites for adsorbates, magnetic properties, and superconductivity.

The differential conductance of the tunnel junction can in principle be obtained by numerical differentiation of the current. However, this method usually suffers from a considerable amount of noise in the current. Only very prominent features in the LDOS will lead to unambiguous changes in σ. Thus, a different method is often used: the lock-in technique[2]. A small oscillating voltage $\Delta V \cos(\omega t)$ is added to the bias, in the ideal case at frequencies far away from high intensity noise, especially above the corner frequency of a typical 1/f behavior in the power spectrum. The current will also acquire an oscillating component at the same frequency ω, which can be retrieved by a special type of amplifier (the lock-in amplifier) even in the presence of strong background noise. By a multiplication of the detected AC signal with its reference (the added voltage) which needs to be compensated by a phase shift ϕ, it is then converted to a DC signal which will finally be low-pass filtered to efficiently attenuate any other unwanted frequency

[2]This method was already used in planar junctions with great success (e.g., [17]) and was later adapted for STM

contribution. It can be shown with the help of the Taylor expansion that this DC signal is then proportional to the first derivative of the current with respect to voltage, i.e., the differential conductance:

$$V(t) = V_0 + \Delta V \cos(\omega t) \tag{2.5a}$$

$$I(t) = I_0 + \sum_{n=1}^{\infty} \frac{1}{n!} \left.\frac{d^n I}{dV^n}\right|_{V_0} (\Delta V \cos(\omega t + \varphi_{\mathrm{n}}))^n \tag{2.5b}$$

$$\begin{aligned} V_{\mathrm{lock-in}} &\propto \frac{1}{T} \int_{t-T}^{t} I(t') \cos(\omega t' + \phi) dt' = \\ &= \underbrace{\frac{I_0}{T} \int \cos(\omega t) dt}_{0} + \frac{\Delta V}{T} \frac{dI}{dV} \int \underbrace{\cos(\omega t + \phi) \cos(\omega t + \varphi_1)}_{(\cos(\phi - \varphi_1) - \cos(2\omega t + \phi + \varphi_1))/2} dt + \ldots \\ &\approx \frac{dI}{dV} \cos(\phi - \varphi_1) \Delta V. \end{aligned} \tag{2.5c}$$

The integration is carried out over a sufficiently long period T (this translates to the aforementioned low-pass filtering), since all oscillating terms will then vanish. The adjustable phase shift ϕ will be tuned to an appropriate value in order to maximize the lock-in signal, compensate an acquired phase, as well as remove the contribution due to the capacitive behavior (Maxwell displacement current) of the tip-sample geometry. In eq. (2.5c) all higher terms with odd n were neglected although they also have a contribution at the fundamental frequency ω, but their amplitude decreases with $1/n!$. Furthermore, the details of the technical implementation (see section 2.1.4) usually lead to a low-pass-like behavior of the whole circuit.

Inelastic Tunneling Spectroscopy (ITS)

In the same way as above, one can use the lock-in technique to extract the nth harmonic from the current by using a reference signal at $n\omega$. However, it rarely makes sense to go beyond the second harmonic, mostly for technical reasons. Then, an analogous formula to eq. (2.5c) can be written down for 2ω which will be approximately proportional to the second derivative d^2I/dV^2. To be useful, it is necessary that the I-V-curve be nonlinear, which is not the case for an ordinary ohmic resistance, but applies for a tunnel junction, although it is not the roughly exponential I-V behavior that is of interest here.

Rather, this technique allows to study a property of solids and adsorbates that has been neglected in the above treatment: their excitation spectrum. The corresponding tunneling process is schematically depicted in fig. 2.3. If one extends the Bardeen formalism (eq. (2.3)) by also allowing tunneling processes that excite the system the matrix element

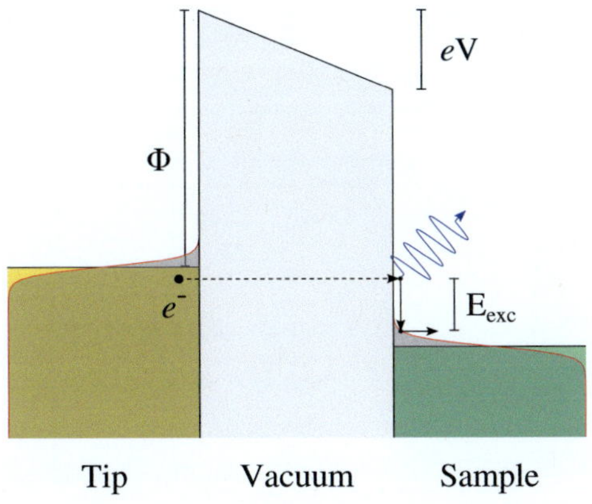

Figure 2.3.: *Inelastic tunneling: During the tunneling process, an electron can transfer energy to an excitation, here symbolized by the blue wiggly line. It must therefore end in a state with lower energy compared to an elastic process.*

M will obtain an additional inelastic term[3]. Then one can show in a similar manner to eq. (2.4) that the second derivative is proportional to the density of states of the excitations D_{exc} (the temperature is taken to be zero for greater clarity):

$$M \rightarrow M' = M_{\mathrm{el}} + M_{\mathrm{inel}} \Rightarrow |M'|^2 = |M_{\mathrm{el}}|^2 + |M_{\mathrm{inel}}|^2 \tag{2.6a}$$

$$I(V_0) \propto \int_{-\infty}^{\infty} dE \Bigg[\rho_1(E)\rho_2(E+eV_0)|M_{\mathrm{el}}(E,V_0)|^2 + \int_{-\infty}^{\infty} dE' \rho_1(E)\rho_2(E+eV_0-E')D_{\mathrm{exc}}(E')|M_{\mathrm{inel}}(E,E',eV_0)|^2 \Bigg]. \tag{2.6b}$$

Since, depending on whether the tunneling was elastic or inelastic, the final states for the tunneling electron are different, there is no interference term between the two processes, and both tunneling probabilities can simply be added (eq. (2.6a)). The approximation that both electronic DOSs $\rho_{1/2}$ can be taken as constant in the voltage range in question leads to the expression

$$\left.\frac{d^2I}{dV^2}\right|_{V_0} \propto D_{\mathrm{exc}}(eV_0)\,|M_{\mathrm{inel}}(eV_0)|^2. \tag{2.7}$$

[3]Such excitations could be phonons, vibration modes in adsorbed molecules, magnons, or spin-flips, for example.

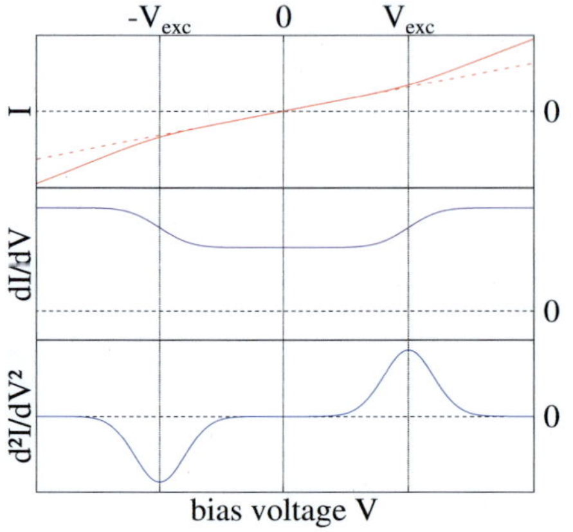

Figure 2.4.: *Idealized expected signals $I(V)$, dI/dV, and d^2I/dV^2 for a single broadened excitation. The red dashed line in the I(V) curve represents the elastic current only.*

The inelastic matrix element also contains information about the probability of the given excitation process, which then should be reflected in the relative signal strength – an additional information that may be used to identify the excitation process if more than one is possible.

As before with STS, this now means that excitation spectra of individual adsorbates, or even local variations of excitations of surface structures can be studied. For many atomic-scale adsorbates like single atoms or clusters thereof, and molecules, the excitation spectrum can be modeled as a sum of delta-like peaks, representing transitions between several discrete states. Therefore, it is instructive to examine the effect of broadening due to temperature and modulation on the energy resolution of this technique. If there was no broadening at all, a single excitation would translate to an infinitely sharp dip-peak pair in the inelastic spectrum, a rectangular step in σ and a kink in the I(V) curve – all occurring at the excitation voltage $|V_{\text{exc}}| = E_{\text{exc}}/e$ (fig. 2.4). Nonzero temperature, however, causes the Fermi function to be broadened, and an added modulation leads to an oscillatory shift of the Fermi levels of the electrodes relative to each other. Both effects can be accounted for by convoluting $D_{\text{exc}}(E)$ with two energy-dependent functions, $\chi(E)$ and $\phi(E)$ [18, 19]. Additionally, the excitation acquires a nonzero linewidth due to coupling to a bath which is accounted for by the scale parameter Γ of a Lorentzian function. In other words, the excited state has a finite lifetime which, by the uncertainty principle, is connected to an energy broadening. For a more practical estimate, χ, ϕ and the intrinsic line shape of the excitation can be approximated by Gaussian functions that each exhibit the same width as the respective original function. This will simplify the convolution since the result is a single Gaussian with a greater effective width W_{tot}:

$$W_{\text{tot}} = \sqrt{\Gamma^2 + (5.4k_BT)^2 + (1.7\Delta V^{\text{rms}})^2}. \tag{2.8}$$

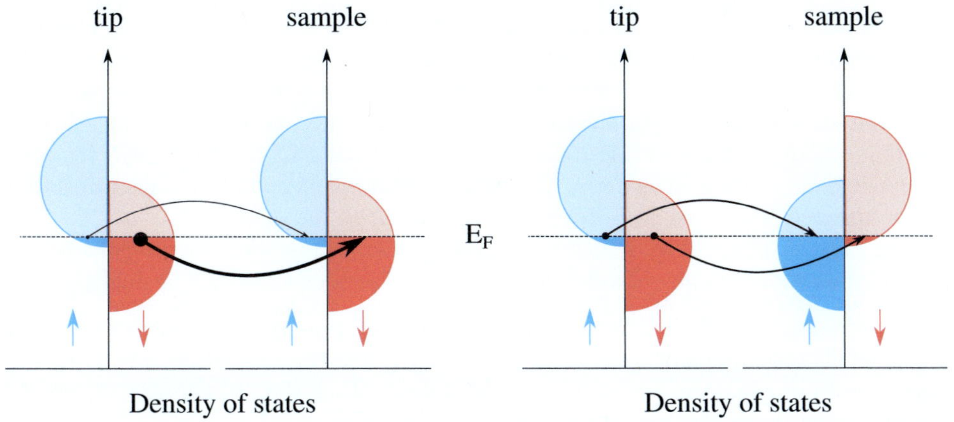

Figure 2.5.: *Schematic illustration of the tunnel magnetoresistance effect: The DOS of a ferromagnet differs for the spin-up (blue) and the spin-down (red) electrons. This leads to a higher current in the case of parallel magnetization, depicted on the left, compared to an antiparallel orientation (right), indicated by the thickness of the black arrows.*

Including these effects, the influence of a single, discrete excitation on the low-energy region of the I(V) curve and its first and second derivative is depicted in fig. 2.4.

Although with this spectroscopy mode one can directly obtain information about the excitation density of states in the second derivative, several reports can be found in the literature that rely only on σ(V) by interpreting conductance steps as (discrete) excitations. One reason for this may be that lower modulation voltages can be used for first derivative spectroscopy. However, minor steps or two steps that partially overlap each other might not be correctly identified, which would drastically alter the interpretation of the observed system.

2.1.3. Spin-polarized STM (sp-STM)

There is one more operational mode that shall be described here which is called spin-polarized scanning tunneling microscopy. It focuses on one specific aspect of the sample's electronic structure: magnetism. The principle behind it relies on the tunnel magnetoresistance (TMR), a magnetoresistive effect that occurs when electrons tunnel through a barrier between magnetic electrodes resulting in a conductance that depends on the relative orientation of the electrodes' magnetization. This effect was first observed by Jullière [20] for ferromagnet-insulator-ferromagnet tunnel junctions and later found its widespread application in modern magnetic hard disks.

Turning back to STM, up to now, the spin degree of freedom of the tunneling electrons was completely ignored in the treatment of the tunnel current[4]. When taking it into account explicitly one needs to define a quantization axis. For STM, the magnetization direction of the tip is usually a suitable choice since it can be assumed that this axis does not unintendedly change during the experiment. As most of this thesis will focus on magnetic atoms with an anisotropy that (supposedly) favors an out-of plane orientation of the magnetization, the z-direction will be chosen as the quantization axis, i.e., the surface normal. The following treatment will be limited to this specific arrangement, yet, a qualitative understanding of different cases should easily be achieved by analogy.

By introducing the different densities of states for the spin-up and spin-down electrons $\rho^{\uparrow/\downarrow}$, and the assumption of spin-conserving tunnel processes, one can already intuitively understand the effect on the tunneling probability. As a simplified example, fig. 2.5 schematically shows the case of two identical, ferromagnetic electrodes, and their spin-split densities of states. There, in the case of parallel magnetization axes, the current can be expressed in analogy to eq. (2.3) as (for $T = 0$)

$$
\begin{aligned}
I(V) &\propto \sum_{\sigma\sigma'} \int_0^{eV} \rho_1^{\sigma}(E)\rho_2^{\sigma'}(E+eV)|M_{\sigma\sigma'}(E,V)|^2 dE \\
&\overset{\delta_{\sigma\sigma'}}{=} \sum_{\sigma} \int_0^{eV} \rho_1^{\sigma}(E)\rho_2^{\sigma}(E+eV)|M(E,V)|^2 dE \\
&= \int_0^{eV} \left(\rho_1^{\uparrow}(E)\rho_2^{\uparrow}(E+eV) + \rho_1^{\downarrow}(E)\rho_2^{\downarrow}(E+eV)\right) |M(E,V)|^2 dE.
\end{aligned}
\tag{2.9}
$$

After introducing the spin polarization η (which can be energy-dependent)

$$
\eta(E) = \frac{\rho^{\uparrow}(E) - \rho^{\downarrow}(E)}{\rho^{\uparrow}(E) + \rho^{\downarrow}(E)}, \tag{2.10}
$$

eq. (2.9) can be rewritten as

$$
I(V) = I_0(V) + C \int_0^{eV} \eta_1(E)\eta_2(E+eV)|M(E,V)|^2 dE, \tag{2.11}
$$

where I_0 denotes an unpolarized current and C accounts for all relevant constant factors in the spin-polarized term. It is evident that, for a contrast to arise, both spin-polarizations need to be nonzero, and that the expected polarized current also depends on the bias voltage.

[4] except as a possible excitation in the case of a spin-flip

In a more general case, when the magnetization of tip and sample cannot easily be assumed to be parallel or antiparallel, the spin-polarized term is again expressed differently. The spin-polarizations are replaced by (local) magnetic DOSs [21] $\vec{m}_{T/S}(\vec{r}, E) = (\rho^{+}_{T/S} - \rho^{-}_{T/S})\hat{e}_{T/S}$ for tip and sample, respectively, where $(+/-)$ denotes the projection of the spin along the magnetization direction $\hat{e}$:

$$I(V) \propto \int_{0}^{eV} [\rho_T \rho_S + \vec{m}_T \cdot \vec{m}_S] \, |M(E,V)|^2 dE. \tag{2.12}$$

Finally, if a constant behavior is assumed for both the unpolarized as well as the spin-polarized part of the tip's DOS, ρ_T and $\vec{m}_T$, one can derive similarly to eq. (2.4) that

$$\left.\frac{dI}{dV}\right|_{V_0} \propto \rho_T \rho_S (E_F + eV_0) + \vec{m}_T \cdot \vec{m}_S (E_F + eV_0). \tag{2.13}$$

Similar to STS, this means that spin-polarized spectroscopy, i.e., the measurement of the differential conductance, can reveal information about the magnitude of the sample's local magnetization projected onto the magnetization axis of the tip. By identifying features in the spectrum that are magnetic in origin, one can then image the surface at this particular bias voltage while simultaneously recording the differential conductance. In this way, local variations of the projected magnetization can be mapped, which is highly useful in revealing the full magnetic structure.

2.1.4. Implementation

Scanning tunneling microscopy belongs to a wider group of techniques, called scanning probe microscopy. All of them have in common that a probing element (e.g., the tip for tunneling) is scanned with sub-nanometer precision across a surface at a very low distance. This is not possible with conventional electromechanic actuators, but only by using piezoelectric elements. Typically, piezoelectric actuators serve two different purposes: First, they form the positioning element for the actual scanning process. Here, various configurations of piezos could be used to make all three dimensions accessible. However, scanner tubes have turned out to be the most common type. They usually achieve lateral scanning ranges of up to several micrometers, while the vertical range is limited to not more than several hundreds of nanometers. Secondly, since it is practically impossible to manually adjust the tip-sample distance such that it is within the vertical range of the scanner tube, another set of so-called coarse motion piezos is implemented. These allow for step-wise large-distance traveling of the tip, or — in a different configuration — a lateral movement of the sample with respect to the tip. Thus, the accessible surface area can be greatly increased.

Another practical challenge is the stability of the tunnel junction: Even if it is possible to position the tip with the above mentioned precision, there need to be extra measures in place to keep this position stable. Hence, appropriate vibration damping is essential which is usually done in several stages with different frequency characteristics. Furthermore, a low-noise voltage supply for the piezo scanner tube is a necessity.

Likewise, the electronics for supplying the bias voltage to the tunnel junction as well as for the detection of the current must meet very high standards. Typically, the tunneling voltage does not exceed a few volts, but high resolution spectroscopy demands very small voltage steps in the microvolt range. Low noise and high fidelity are the fundamental requirements here. Especially in low temperature instruments the electronic temperature may differ from the thermodynamic temperature. Thus, the bias voltage is often subject to additional low-pass filters in order to reduce any radio frequencies (and higher) that might be picked up by the cables.

Typical tunnel currents range from several picoamperes to well above 100 nA. The current has to be translated into a voltage and properly amplified for measurement and further signal processing. The amplification by so-called transimpedance amplifiers[5], however, involves a tradeoff because effectively the amplifiers also act as a low-pass filter: higher gain — which means better signal-to-noise ratio — usually lowers the bandwidth of the output signal. This bandwidth normally lies in the kilohertz range and can thus have a marked influence on the feedback loop, or on the parameters used for spectroscopy.

2.2. Experimental Setup

The measurements presented in this thesis were performed with a homebuilt low temperature STM, the Joule-Thomson STM (JT-STM), which shall be portrayed in this section. Some figures concerning the performance of the instrument will be given.

This particular STM had been designed with the aim (among others) of investigating magnetic properties of nanoscale systems. Therefore, it is equipped with a magnet made of superconducting wire. Another STM (called "dilution STM" or d-STM) which was built in the group of Prof. Wulfhekel was designed to feature such a magnet as well. A short section at the end of this chapter presents this d-STM magnet, since the work of finalizing the design as well as manufacturing and testing the magnet was a part of this thesis.

2.2.1. The Joule-Thomson STM

The "Joule-Thomson STM", developed by Zhang, Dehm, and Wulfhekel [22, 23], constitutes not only the actual microscope itself and the related cryostats, but a complete

[5] also called pre-amplifiers or I-V-converters

facility for *in situ* preparation of tips and samples.[6] The supporting frame resides on four pneumatic vibration isolator feet (Newport S-2000 series) to decouple the system from vibrations transmitted through the floor.

It consists of three connected ultra-high vacuum (UHV) chambers – the STM chamber, a preparation chamber, and a small load lock — between which objects can be moved by transfer rods. All chambers share a common roughing pump and are equipped with their own turbomolecular pumps as well as standard Bayard-Alpert ionization gauges for pressure measurement. Furthermore, there is an ion pump in each of the two major chambers to sustain UHV conditions, so that the turbomolecular pumps, which usually produce vibrations, can be shut down during the actual measurements. In the following, those details of each chamber that are relevant for this thesis will be mentioned.

STM Chamber

As was stated in eq. (2.8), the energy resolution during tunneling spectroscopy depends critically on temperature, as well as modulation voltage. While vibrational energies of some adsorbates (such as simple molecules) can easily exceed the energy resolution of a room temperature measurement [5], spin excitations tend to be on the scale of a few millielectronvolts [6]. For this reason, a Joule-Thomson (JT) expansion stage for helium gas supplements two conventional bath cryostats — one for liquid nitrogen (20 l), the other for liquid helium (10 l) — to achieve sub-Kelvin temperatures. Initially it was used with pure ^{4}He and the lowest possible temperature was approximately 900 mK; after replacing the gas with a ^{3}He-^{4}He-mixture, the lowest achievable temperature now amounts to 650 mK in single-shot mode and around 700 mK under continuous operation.

Among the notable characteristics of the cooling system is its low consumption of cryogenic liquids, especially liquid helium: Under ideal conditions it can last for up to 280 hours, which allows for long periods of uninterrupted measurement time.

Furthermore, the cryostats act as cryopumps that help to achieve a pressure in the low 10^{-10} mbar range in the volume surrounding the outermost radiation shield. The sample and tip reside inside these shields (three in total) and are accessed through small sliding doors in the shields. There, although it cannot be measured, the pressure is even better which further reduces sample contamination by adsorbates from the residual gas.

The chamber is equipped with an electron beam evaporator which points at the sample already in its measurement position. This way, single atoms that tend to diffuse across surfaces at elevated temperatures can be deposited at less than 20 K to ensure that most of the atoms remain where they hit the surface. During deposition, the pressure rarely

[6]Its design was licensed by SPECS GmbH and a derived instrument is commercially available; see, e.g., http://www.specs-zurich.com

rises above 10^{-9} mbar, so that in those short time intervals (usually less than a minute) only a negligible contamination of the sample surface can occur.

Finally, a 3 T split-coil magnet made from superconducting wire completes the setup, allowing to apply fields perpendicular to the surface.

Preparation Chamber

As the name suggests, almost all of the preparation steps take place in this UHV chamber. This includes standard sample or tip sputtering with argon ions, usually at 3 kV, annealing and flashing of substrates, as well as deposition of (mostly metallic) materials onto the prepared substrates and tips (e.g., for sp-STM, see section 2.1.3). The chamber features a titanium sublimation pump which, together with the ion pump, helps to keep the pressure in the low 10^{-10} mbar range.

The ion gun is equipped with two leak valves, one for standard argon sputtering, the other for nitrogen in order to prepare copper nitride surfaces (see chapter 5). Being a differentially pumped ion source, it limits the pressure increase to the low 10^{-8} mbar range in typical sputtering conditions.

Additional equipment consists in a combined unit for LEED and Auger spectroscopy to analyze the geometric and chemical properties of the surface by scattering of electrons, a quadrupole mass spectrometer to determine the residual gas composition, a pyrometer to measure the sample temperature (range: 250 °C to 800 °C), and several electron beam evaporators. The relevant evaporators are loaded with cobalt (Co), and either chromium (Cr) or manganese (Mn). The latter two metals were used to coat the tips with antiferromagnetic material for sp-STM while cobalt was occasionally used as a reference material to confirm the nonzero spin-polarization of the tip (see section 4.2).

Typically, the preparation of platinum (111) crystals consisted in repeated cycles of Ar ion sputtering at 3 kV and subsequent annealing between 550 °C and 800 °C. After the final cycle, the sample was either transferred to the STM, or subject to cobalt deposition of less than ca. 0.5 monolayers (ML). Similarly, newly etched tungsten tips were degassed, sputtered and flashed several times to remove any remaining oxides at the tip apex. For sp-STM, 20 to 40 ML of chromium (following [24]) or about 10 ML of manganese were deposited onto the tip.

Load Lock

This chamber is the smallest of the three chambers and serves as the main entrance and exit for samples, tips, or other utilities mounted on sample plates. It also provides the pumping line for the differentially pumped ion gun of the preparation chamber. Since it is vented much more often, compared to the two main chambers, the pressure is of lesser importance and normally ranges between 10^{-9} and 10^{-8} mbar.

gain (V/A)	10^9	10^8	10^7
bandwidth (kHz)	1	7	50

Table 2.1.: *Gain-dependent bandwidth of the Femto DLPCA 200 I-V converter. For very low currents, when the highest gain (10^9) should be used, the low bandwidth severely affects the usable frequency range for the spectroscopy modes, even more so when considering that the second derivative spectroscopy makes use of the second harmonic.*

Electronics and Instrumentation

Finally, a short overview of the electronics used for this setup shall be given. The central part is a stack of Nanonis SPM control components, namely a real-time controller (RC4), an 8-input/8-output signal conditioner (SC4), as well as three cooperating units that provide the high-voltage for the coarse motion piezo elements and the scanner tube (HVA4, PMD4, HVS4).

A computer runs the Nanonis SPM Control Software V4 that communicates with the RC4. Although this program also offers a software lock-in module, currently the JT-STM setup uses two external analog dual-channel lock-in amplifiers (EG&G Instruments 5210) to be able to conduct first and second derivative measurements simultaneously. The oscillator output of one of these amplifiers is connected to the bias voltage via a Nanonis SA4 signal adder.

Before the measurement of the tunnel current can happen, it needs to be amplified (see section 2.1.4). For this a Femto DLPCA 200 pre-amplifier is used. The typical gain settings are "low noise", 10^9 to 10^7 V/A, which have different bandwidths (at this frequency the attenuation amounts to -3 dB), given in table 2.1.

Lastly, the superconducting coil is driven by a Oxford instruments Mercury iPS-M magnet power supply which can provide up to ± 60 A, though this magnet reaches its maximum field at 35 A.

2.2.2. A Superconducting Solenoid for the Dilution STM

The d-STM is a newly built facility, in its general concept not quite unlike the JT-STM, although a much greater attention was paid to build it as modular as possible. This characteristic trait facilitates all maintenance activity, but also enables one to substitute parts with alternative implementations. Therefore, the magnet that will be described in the following paragraphs is only one possible option – if the need should arise, other configurations (e.g., for vectorial fields) may be fabricated and installed.

From the beginning the d-STM was conceived with a superconducting magnet in mind that ought to surpass the performance of the JT-STM type significantly. Again, the demand was for a magnetic field that is oriented perpendicular to the sample surface. To achieve this, the general concept was to build a nested solenoid, submerged in liquid helium (also called a "wet" magnet), with its subcoils connected in series. While the coil should be made from "conventional" low-temperature superconductor (NbTi) wires, the current leads were planned as two segments – a high-temperature (YBCO) superconductor tape between the coil and a contact at roughly 77 K, and a copper wire of appropriate thickness from there to the feedthrough at room temperature. In this design, there was also no plan for a persistent switch, although one might still be retrofitted if needed.

During this thesis, the attempt was made to find the best solution under given constraints, such as limited size or the incentive to use the lowest possible current in order to reduce resistive heating of the non-superconducting parts. Since there existed already a draft for a vessel to house the coil and liquid helium, the task left for this thesis focused on the construction of the coil itself. After that the coil was manufactured at both the mechanic and electric workshop of the Institute of Physics at KIT and subsequently tested. Several people shall be mentioned here who contributed to this project, starting with Prof. W. Wulfhekel who conceived of the magnet as a whole, Dr. Lei Zhang who created the initial CAD drawings, and Dominik Rau, who performed much of the actual manufacturing work.

The Optimization Problem

Since there exists no analytic formula to calculate the magnetic field distribution of a finite-sized coil, one has to resort to reasonable approximations or numerical solutions. Likewise, there is no general recipe to find the best-performing nested-coil geometry, one usually has to iteratively improve and evaluate the intermediate solutions. In order to elucidate this procedure, the essential aspects of the available design degrees of freedom will be described in the following.

Before any of the ingredients will be covered the dimensions of the coil shall be touched on quickly since the overall size was fixed: The axial length of the winding volume is 90 mm, the innermost winding has a radius of 24 mm, and the outer radius of the outermost subcoil is 70 mm. The bobbins of the subcoils were planned to be made of antiferromagnetic steel, with a wall thickness of 1 mm, slightly reducing the available winding volume.

diameter (µm)				
bare	insulated	ratio Cu:SC	type	shorthand notation
300	330/340[a]	1.3 : 1	54S-43	I
400	440	1.3 : 1	54S-43	II
500	540	1.3 : 1	54S-43	IIIa
500	540	0.9 : 1	56S-53	IIIb
600	640	1.3 : 1	54S-43	IV
700	750	1.3 : 1	54S-43	V

Table 2.2.: *Overview of the wires in question. Almost all wires have a 1.3:1 ratio of copper matrix to superconductor material, except for one of the* 0.5 mm *wires. A shorthand notation is introduced and will be used for the remainder of this chapter.*

As stated above, standard niobium-titanium wires were used. The supplier of choice was Supercon[7], which offers a wide variety of NbTi wires with a copper matrix. Two empirical factors influenced the choice of wires: First, the more ampere-turns one can fit in the available volume the higher the achievable field can be — a fact that suggests taking thinner wires. Secondly, the current was expected to be of the order of a few tens of amperes, which establishes a minimal wire diameter, given that the ability of a wire to carry a supercurrent decreases in magnetic fields. Other practical issues such as mechanical stability and winding effort played a minor role, as well. The wires that were considered are given in table 2.2, and their performance in terms of field-dependent critical current $I_C(B)$ at 4.2 K is plotted in fig. 2.6. Although the numbers might seem impressive (e.g., ~100 A at 9 T for wire V), one has to keep in mind that the obtainable performance is usually considerably lower. According to [25, 26] a maximum current of 50-70% of I_C at a given field is a reasonable estimate. Thus, for all calculations a conservative limit was used, meaning that the current should never exceed 50% of a wire's lowest I_C which, in turn, is determined by the coil's own field.

The three major questions that had to be answered were:

- Into how many subcoils should the magnet be divided?
- Which combination of wires gives the highest magnetic field?
- What is the optimal geometry of the subcoils?

All of these questions are, however, interconnected (e.g., the optimal geometry cannot be the same for two different combinations of wires) and thus the answers were found by direct comparison of the performance of various designs.

[7] www.supercon-wire.com

[a] Different values were found online and on the data sheet that came with the shipment. Some variance should be expected.

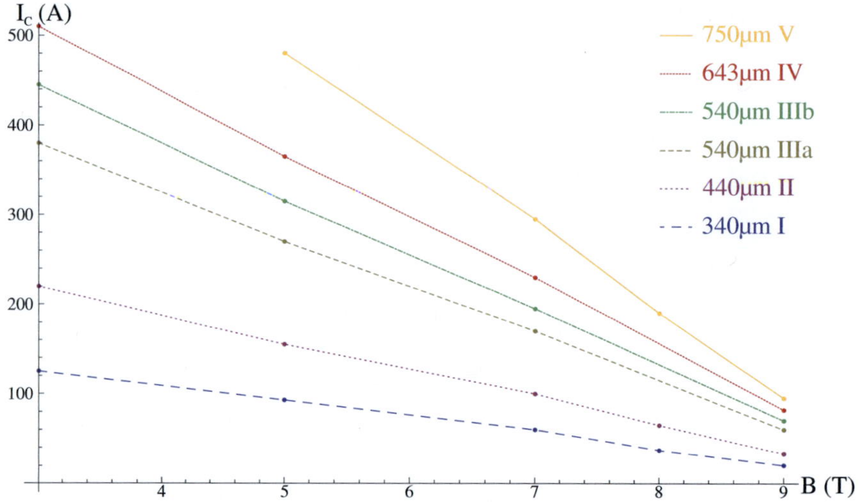

Figure 2.6.: *Behavior of the wires' critical currents I_C with an externally applied field B. Data points were taken from the certificate of compliance that came with the shipment, or from [27].*

To this end, two complementary computer programs were used: On one hand, Mathematica [28] code was written to implement the calculation of B_z at the points of interest and to subsequently find the optimal geometry for a given set of wires. In this calculation, the data presented in fig. 2.6 was linearly approximated up to 9 T, but no prediction was made for higher fields. On the other hand, the freely available finite element solver FEMM [29] was utilized to visualize the two-dimensional field distribution, to extract characteristic values of the whole assembly, such as the inductance L or the total magnetic energy, and for consistency checks.

From the beginning, it should be clear that the outermost coil would be made of the thinnest wire, and that in the segments towards the center the wires need to become increasingly thicker. Hence the number of possible combinations was not as high as might be expected. Likewise, it became clear quite early that a two-coil configuration would always be inferior to a similar three-coil version. Due to computational costs and again higher manufacturing efforts, a four-coil option was not considered, and so several variants of the three-coil configuration were examined.

Secondly, a general trend was observed, concerning the wire thickness: Compared to the thinnest usable wire in the innermost coil (IIIa/b), a thicker wire would always perform worse – either delivering lower fields or demanding higher currents. Thus, only the wires I-IIIa/b were left. The decision not to use wire IIIb despite its better superconducting properties was a decision in favor of safety – the improved I_C comes at the expense of a sufficiently thick copper matrix that could briefly conduct the applied current without breaking in case of a magnet quench.

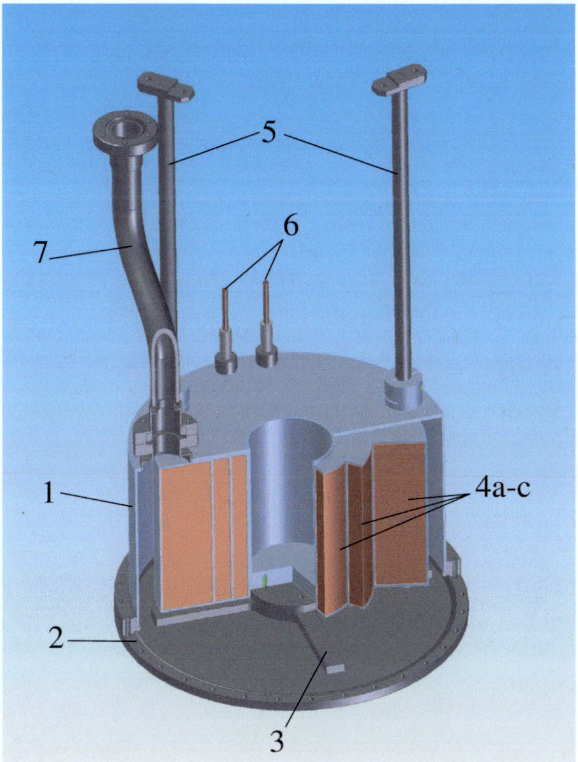

Figure 2.7.: *Cross-sectional CAD view on the components of the magnet, showing the outer vessel (1) and its main flange (2), a cross-shaped steel mount (3) for the bobbin where the three subcoils (4a-c) and their soldered joints are fixed (not shown), two of three steel legs (5) for mounting the whole assembly to the bottom of the helium cryostat, copper contacts for the current (6), and a flexible steel tube with flanges (7) to connect the coil vessel with the liquid helium tank.*

The final step of finding the best subcoil radii, in due consideration of the 50% rule and using the lowest possible current, favored rather thin inner coils of 8 mm (inner) and 9 mm (middle) thickness. This leaves the outer coil 27 mm thick. The expected performance would then amount to B_z=8.52 T in the geometric center of the coil at a current of 32.45 A. A comparison of all characteristic figures of the calculation and the achieved values after winding will be presented at the end of the next section in table 2.3.

Figure 2.7 offers a cross-sectional CAD drawing of the complete assembly which, in turn, was installed in the d-STM.

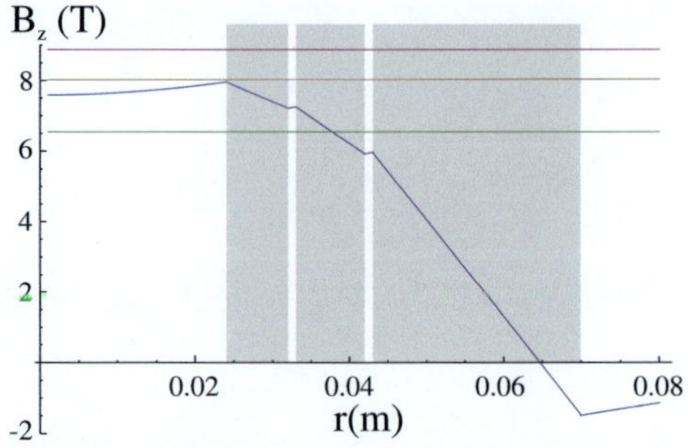

Figure 2.8.: *Calculated B_z as a function of the radius r in the central plane (z=0) for the finished magnet at 32.25 A. At the sample position the field amounts to 7.57 T. The three horizontal lines represent 50% of the lowest critical current of the wires IIIa, II, and I (from top to bottom), respectively, evaluated at the inner radius of each subcoil. Obviously, there is still a small margin, possibly allowing slightly higher currents before reaching 0.5 I_C. The shaded areas mark the volumes of the subcoils.*

Fabrication and Testing

After the various coil parts had been made by the mechanical workshop of the Institute of Physics the winding process began. During this fabrication step the wires were impregnated with a two-component epoxy glue[8] that was specifically chosen for its beneficial low temperature properties. Its purpose is mainly to embed the wires in a rigid, though not brittle, matrix to restrict the wires from moving due to Lorentz forces in high fields which could cause a quench of the magnet. Moreover, the epoxy provides additional electrical insulation and heat conductance.

It should be noted that the winding did not go without problems. A few times, it could not be avoided that the wire, or at least its insulation, was slightly damaged, possibly deteriorating the wire's performance locally. Furthermore, it had to be accepted that the estimated total number of windings could not be achieved for any of the three subcoils and that the assumptions that led to these numbers were too optimistic. In a final step, when the three subcoils were assembled into one unit, a minor electrical short to ground was found, meaning that at least one wire's insulation was scraped off during this step, with the possibility that also the wire was damaged. Although it must have been one of the outermost wires where the magnetic fields will never be close to the critical values, this might still restrict the coil from achieving its full potential.

With the winding numbers of the subcoils known, the calculations were updated to give a new estimate of the magnetic field. The results can be found in table 2.3 as well as in fig. 2.8 and fig. 2.9.

The last stage of the whole fabrication process was, of course, to test the finished coil. To this end, an unused cryostat for liquid helium was reactivated. The coil was then

[8]EP29LPSPAO-1BLACKZ0010 by Master Bond Inc.

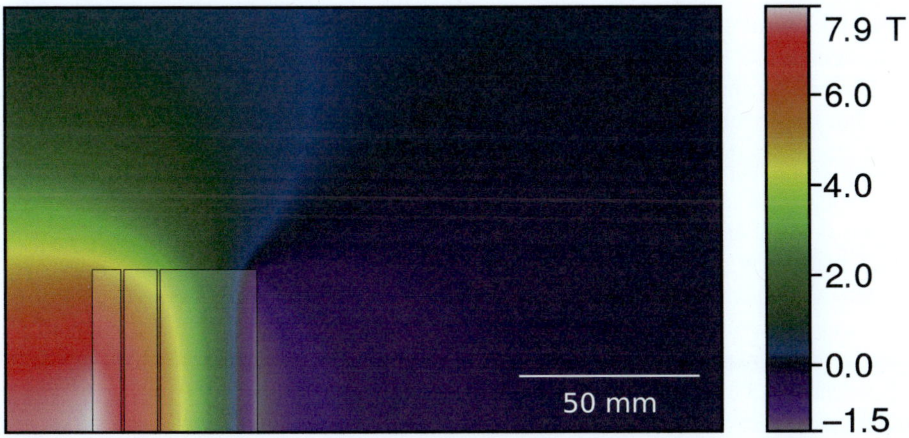

Figure 2.9.: *Color plot of the finite element calculation of B_z in the upper half of the $r-z$ plane. The calculation is based on the winding numbers of the finished coil, and shows the situation for a current of* 32.25 A. *In the lower left corner the shaded regions represent the winding cross section area. The highest field values can be found in the central plane (z=0) at the innermost windings of the subcoils. It is the field there that ultimately limits the magnet.*

installed, connected and cooled down to 4.2 K. Two test runs were conducted: The first run, performed at a ramp rate of 0.5 A min^{-1}, reached 26.7 A which translates to 6.27 T. The second test run slightly surpassed the first one, reaching 27.5 A which amounts to 6.46 T. This can be considered a training effect [26], but might as well be due to a reduced ramp rate (0.2 A min^{-1} above 20 A).

After these two tests no further training was attempted since the performance of the coil was already satisfying for the intended purpose. Possibly, even higher field values could have been achieved, since more training effects can still be expected. Instead, the coil was finally installed in the d-STM, and tested there up to 5 T, while carefully monitoring the temperature of the various superconductor contacts (coil-to-high-TC, and high-TC-to-copper leads). None of the temperatures increased by more than 50 mK in the course of more than 1.5 hours. The coil and the d-STM were then declared fully operational.

Conclusion

Although some difficulties were experienced during the manufacturing process, at the end a sufficiently well performing coil was produced. Several lessons were learned about the validity of the assumptions for designing the coil geometry, and about the manufacturing process. The initially assumed hexagonal stacking (filling factor 90.7%) for the thickest wire could not be achieved in the respective coils. An almost perfect quadratic stacking can be obtained for these relatively flat coils, however, the available thickness did not correspond to an even number of layers and thus, some space could not be used. The approximation of a wild winding with the same filling factor as a quadratic stacking (78.5%) can be considered moderately accurate, though 70% is closer to reality in this case. For future fabrications, even more care has to be taken in order not to damage any wires, especially during the process of combining the subcoils.

Some attempts were made to measure the actual field of the coil. However, the available sensors were either not adequate (their field range was far too small), or the readings were not accurate enough. This last fact is also due to some of the coil's parts being weakly magnetizable, though most of the assembly parts are made from antiferromagnetic steel. Some welding seams and screws, however, can be magnetized and exhibit a remanence. These measurements were done at room temperature since no low temperature sensor was available. Thus, the currents that could be applied were far from surpassing the saturation field of the magnetizable components. This evaluation of the coil's true $\vec{B}(I)$-characteristic is left for the future. For new constructions of coils one should employ non-magnetic materials for all bobbins, mounts, screws, or housing.

		calculated	as-built	discrepancy (%)
number of windings	I	2851/2324	2238	-21.5/-4.4
	II	4184/4080	3774	-9.8/-7.5
	IIIa	21020/20800	19204	-8.6/-7.7
filling factor (%)	I	90.7/73.9	71.2	
	II	78.5/76.6	70.8	
	IIIa	78.5/75.7	71.8	
maximum B_z(r=0) in Tesla		8.52/8.21	7.57	-12.5/-7.8
conversion factor I/B_z (A T^{-1})		3.81	4.26	-
Inductance L in Henry		41.7	34.35	-

Table 2.3.: *Characteristics of the finished coil. The two possible theoretical winding numbers in the "calculated" column are due to different assumptions about the stacking density of the wires. The first number relies on an oversimplified modeling, dividing the available winding cross section by an effective area per wire. The second value is a more realistic approach, taking into account the number of windings per layer and the number of possible layers, as well as a reduced stacking efficiency, equal to a quadratic arrangement of the wires. The values for B_z, I/B_z, and L are all calculated and have not been verified by a measurement, yet.*

3 Chapter 3.

Magnetism of Single Atoms on Surfaces

As was stated in the introduction, this thesis is the continuation of existing work by Schuh et al. [10, 30]. A model to describe a single magnetic atom on a surface was developed in those works, the most notable trait being that both spin and orbital contributions to magnetism were unified by using the total angular momentum $\vec{J}$. This allows one to construct a crystal field Hamiltonian that incorporates the full symmetry of the adsorption site of the atom. Its application specifically to holmium atoms on a platinum-(111) surface was later successfully used to explain the experimental discovery of surprisingly long lifetimes of the magnetic ground states [9]. The fundamentals, details, and limitations of this model will be discussed in this chapter.

3.1. The Crystal Field Hamiltonian

3.1.1. Introductory Notes

It is a well-known fact that the key ingredient to obtain preferred magnetization directions is having magnetic anisotropy. This anisotropy results from breaking the rotational symmetry of a free atom. While all states of one multiplet are degenerate in the free atom they experience a zero-field splitting (ZFS) in the presence of matter in its environment (be it atoms of a bulk crystal or surface, or ligands in molecules), that partially lifts the degeneracy and leads to different energy eigenvalues. The mechanism for this splitting is two-fold: First, the proximity of other atoms or ions gives rise to an anisotropic Coulomb potential, called the crystal field (CF). Secondly, the intra-atomic spin-orbit interaction couples the affected orbital angular momentum states to the spin degree of freedom. Thus it is clear that neither m_L nor m_S, i.e., the eigenvalues of the operators[1] L_z and S_z, can be good quantum numbers anymore[2]. Rather, one has to use the total angular momentum $\vec{J}$ and its components J_x, J_y, J_z. It is assumed here, that $\vec{J}$ is defined to describe the whole impurity: magnetic atom and the potentially induced substrate polarization. Then $\vec{J}^2$ and J_z must obey the known quantum mechanical rules for angular-momentum type operators. Their quantum numbers will be denoted as J and M, the J_z-expectation value as $\langle J_z \rangle$. A few remarks should be given before elaborating on the details of the mathematical description:

- The remainder of the single atom topic will focus on the lowest (= ground state) multiplet only. This is because all experiments were performed at sufficiently low temperatures, and the energy differences within one multiplet are much smaller than those between the two lowest multiplets.
- A good start to estimate the ground state multiplet are Hund's rules, which are briefly stated in the appendix A. In the case of holmium atoms on Pt(111), it was later confirmed by *ab initio* calculations that their use was appropriate.
- It will be introduced in the next section how the crystal field can be expressed in terms of the total angular momentum. This relies on the Wigner-Eckart theorem and involves the calculation of coefficients which depend on J. Hence, each J-multiplet must be described with its own set of crystal field parameters.
- An external magnetic field $\vec{B}$ has an influence on the symmetry properties of the system, most notably the time inversion invariance. For the time being, the field is assumed to be zero. The discussion will later include its effect on the spectrum and dynamics.

3.1.2. Stevens Operators

When considering the effect of adsorption of a single atom onto a non-magnetic metallic surface, the first step to describe the system is to find an appropriate formulation of the crystal field Hamiltonian $\mathcal{H}_{CF}$. Although this could be done in several different ways, the most useful involves the so-called Stevens operators [31, 32]. They will be denoted as O_n^m, and will be accompanied by their respective anisotropy constants B_n^m where both n and m are integers. These operators are n^{th} degree polynomials composed of the operators $\vec{J}^2$, J_z, and the ladder operators $J_\pm = (J_x \pm iJ_y)$ which are raised to the mth power.

The beauty of this formulation becomes apparent when one considers the properties of the adsorbed atom and their relevance for the B_n^m: The index n takes into account the total angular momentum properties of the atomic shell. It has to be a nonnegative, even

[1] All orbital-momentum like operators are to be understood with $\hbar = 1$; hence, their eigenvalues and expectation values are dimensionless numbers.

[2] For atoms on a surface the z-axis will be chosen parallel to the surface normal.

number due to time inversion symmetry, and is limited by $n \leqslant 2J$ or $n \leqslant 2\ell$ (whichever applies first), where ℓ is the single-electron orbital angular momentum quantum number (in the case of rare-earth elements, $\ell = 3$, for transition metals $\ell = 2$). Similar rules can be stated for m: The point symmetry group of the adsorption site directly translates to the nonzero terms, i.e., an m-fold rotational symmetry allows only those operators $O_n^{m'}$ that have $m' = km, k = 0, 1, 2, \ldots$ All other coefficients must vanish. Lastly, its range is limited by $0 \leqslant m \leqslant n$. Thus, from the symmetry and the knowledge of J, one can easily set-up the general structure of the full crystal field Hamiltonian:

$$\mathcal{H}_{\mathrm{CF}} = \sum_{n=0}^{\infty} \left[B_n^0 O_n^0 + \sum_{m=1}^{n} \left(B_n^m O_n^m + \tilde{B}_n^m \tilde{O}_n^m \right) \right]. \tag{3.1}$$

The terms $\tilde{B}_n^m \tilde{O}_n^m$ can be made to vanish completely if the adsorption site symmetry also features a mirror plane normal to the surface. All operators of interest for this thesis, i.e., up to $n = 6$ and the allowed operators for a given symmetry can be found in the appendix (eq. A.1, A.2, A.3, and table A.1).

What is not a priori known, and difficult to obtain, are the coefficients B_n^m. To date, there is no sufficiently accurate way to analytically calculate them from known properties of the atomic shell and substrate. Thus, they must be deduced with the help *ab initio* calculations or be estimated from measurements.

3.1.3. The Spectrum of $\mathcal{H}_{\mathrm{CF}}$

In order to distinguish between different sets of states, the following notation will be used: Since all states are eigenstates of $\vec{J}^2$ with the same eigenvalue the quantum number J will be omitted in the notation without exception. Eigenstates of J_z will be denoted by $|M\rangle$, while $|\Psi_M\rangle$ represents the CF-Hamiltonian eigenstates. In the latter, the subscript M signifies which of the J_z-eigenstates has/have the largest contribution(s). Where necessary, an additional superscript, a or s, will denote antisymmetric or symmetric linear combinations of degenerate states with the same $|M|$. Note that the $|M\rangle$ and $|\Psi_M\rangle$ can be identical under certain conditions.

By considering the effect on the spectrum, the Stevens operators can be divided into two groups – operators with $m = 0$ and those with $m \neq 0$. The former contain only powers of $\vec{J}^2$ and J_z; therefore, if no other operators are present, the eigenstates of the CF-Hamiltonian are still the set of $|M\rangle$ states. If, however, terms with $m \neq 0$ appear in $\mathcal{H}_{\mathrm{CF}}$ they contain the operators $J_\pm^m$ which lead to mixing of different $|M\rangle$ that form the new eigenstates. The solution for an m-fold symmetric system is then the set of $|\Psi_M\rangle$:

$$|\Psi_M\rangle = \sum_{M'=-J}^{J} c_{\mathrm{M,M'}} \left|M'\right\rangle. \tag{3.2}$$

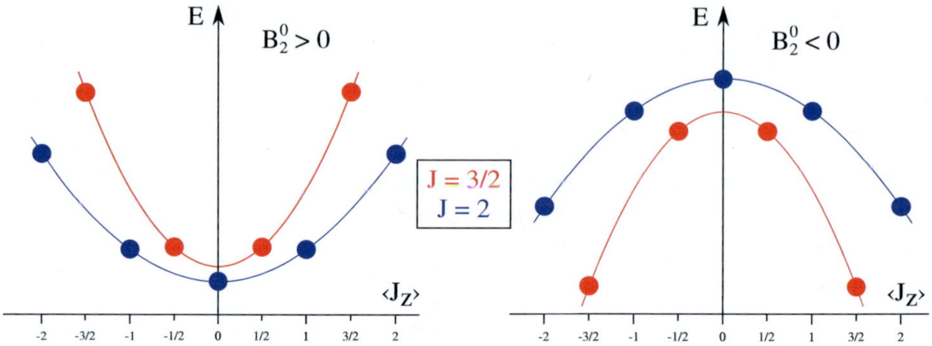

Figure 3.1.: *Effect of the operator O_2^0 on the spectrum of integer and half-integer systems, represented by $J = 2$ and $J = {}^3\!/\!_2$, respectively. The general shape of the spectrum is shown for both easy and hard axis. On the left the diagram illustrates the different degeneracy of the ground state(s), whereas on the right both systems exhibit doubly degenerate ground states.*

For $m > 1$ some of the $c_{\mathrm{M,M'}}$ vanish; only those coefficients with $M' = M + km$ are nonzero, where k is an integer number.

A few examples shall illustrate the influence of the most important operators. Since n must be even, the lowest order operator can only be O_2^0. It describes the uniaxial anisotropy in the direction of the surface normal:

$$O_2^0 = 3J_z^2 - \vec{J}^2 \tag{3.3}$$

If only O_2^0 is present the adatom has an out-of plane easy or hard axis for its magnetization, depending on the sign of the coefficient B_2^0. It is also important to note at this stage already that it makes a difference whether J is a half-integer or integer number. As the examples of $J = 2$ and $J = {}^3\!/\!_2$ in fig. 3.1 illustrate, there are two ground states in three out of four combinations. Only for integer J with $B_2^0 > 0$ the ground state has $\langle J_z \rangle = 0$ and is non-degenerate, which is not desirable for the application as a magnetic bit.

If one allows an in-plane anisotropy, the lowest nontrivial symmetry is a two-fold rotation which introduces the operator O_2^2, see eq. (3.4). A two-fold symmetric adsorption site may allow further terms with higher n and $m = 2$ that affect the spectrum in almost the same way as O_2^2 so that they will not be discussed separately.

$$O_2^2 = \frac{1}{2}\left(J_+^2 + J_-^2\right). \tag{3.4}$$

Obviously, O_2^2 couples $|M\rangle$ that differ by two so that the resulting $|\Psi_M\rangle$ states are no longer eigenstates of J_z. The effect on the spectrum is shown in fig. 3.2 and accentuates the dramatically different behavior of half-integer vs. integer J systems. In a half-integer system $\langle J_z \rangle$ is slightly decreased for all states and their energies can shift. These minute changes are in stark contrast to the integer J situation: There, even a weak B_2^2 (weak

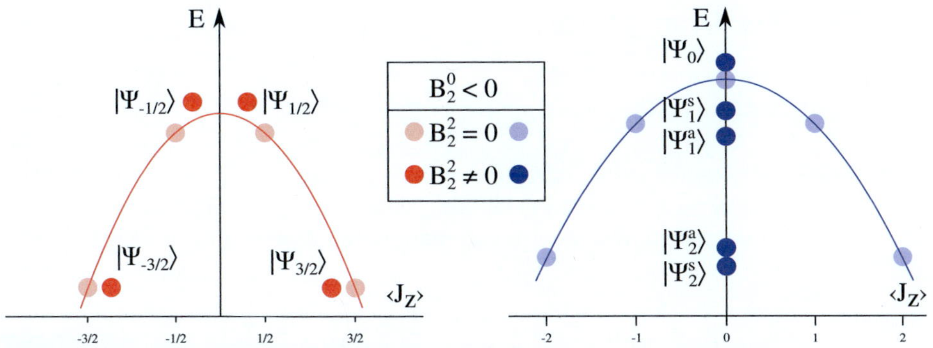

Figure 3.2.: *Consequences of a nonzero B_2^2 in an out-of plane easy-axis scenario. The spectrum of a half-integer J system does not change much, but in the integer J system, all states lose their magnetic moment, and the formerly degenerate $|M\rangle$ states form symmetric and antisymmetric pairs, splitting in energy. (It is assumed that $B_2^0 < 0$; the opposite situation yields the analogous result.)*

enough so that O_2^0 still dominates) couples all states with equal $|M|$ so that $\langle J_z \rangle = 0$ for the complete spectrum. In particular, this coupling leads to (anti)symmetric combinations[3] of $|M\rangle$ states with equal $|M|$, resulting in the energy-split $\left|\Psi_M^{a/s}\right\rangle$ (see also [33]).

This is more than just coincidence, but it goes back to fundamentally different properties of half-integer and integer J systems, summarized by Kramers theorem. It states that the spectrum of a half-integer system, i.e., one with an odd number of electrons, will always remain at least two-fold degenerate, due to time-inversion symmetry. Ultimately, this goes back to a different transformation of half-integer spins with respect to the time inversion operator $\mathcal{T}$ (see section 3.3), compared to integer spin. The spectrum of a system with an even number of electrons and hence an integer J is not protected in such a way and can have non-degenerate states at $\langle J_z \rangle = 0$.

The bottom line of these examples is that, from a few symmetry considerations, one can already make some assertions about the properties of a system. Integer J will not lead to magnetic states at all in a two-fold symmetric environment. It is also obvious that all operators have an influence on the excitation energies – in some cases it is limited to a mere shift, in other cases the contribution decisively influences the arrangement of the states, and thus defines the ground state and excited states.

[3] Whether the lower state is the symmetric or antisymmetric combination can depend on the sign of B_2^2.

3.2. Further Interactions

3.2.1. Magnetic Field

So far, the only interaction to be considered was the crystal field of the substrate. There are, however, other aspects that need to be taken into account. One important additional term for a full Hamiltonian is due the Zeeman energy of a magnetic moment in an external magnetic field $\vec{B}$:

$$\mathcal{H}_\mathrm{Z} = -g_\mathrm{J}\mu_\mathrm{B}\vec{J}\cdot\vec{B}, \text{ with } g_\mathrm{J} = 1 + \frac{J(J+1) - L(L+1) + S(S+1)}{2J(J+1)}, \tag{3.5}$$

where μ_B denotes the Bohr magneton, and g_J is the Landé g-factor. The g-factor of an electron's spin is approximated as $g \approx 2$. Due to the possibly high number of parameters that are needed to describe an adatom (J, CF parameters, temperature, $B_{\mathrm{x,y,z}}$, ...) it is not possible to make a general statement about the influence of the magnetic field on the spectrum. A few special cases, however, can be considered.

If the field is perpendicular to the surface ($\vec{B} = (0, 0, B_\mathrm{z})$) the rotational symmetry is preserved and the effect of the field can be evaluated systematically. Then, the field lifts the aforementioned degeneracy, both in integer and half-integer systems and tilts the parabolas shown in fig. 3.1. If B_z is strong enough, it can also counteract the splitting induced by the Stevens operators with $m > 0$ (fig. 3.2) and restore a nonzero $\langle J_\mathrm{z}\rangle$ for the affected states.

For an additional in-plane field component, the situation is different. $\vec{B}_{\mathrm{xy}}$ disturbs the spatial symmetry and effectively reduces it to a $C_{1(\mathrm{v})}$ symmetry — mathematically speaking, the additional operators $J_\pm$ appear in the full Hamiltonian $\mathcal{H}_\mathrm{total} = \mathcal{H}_\mathrm{CF} + \mathcal{H}_\mathrm{Z}$, coupling all $|M\rangle$ with each other. How this influences the composition of the eigenstates $\mathcal{H}_\mathrm{total}$ must be calculated for each individual case.

However, as long as B_z is of similar strength as $|\vec{B}_{\mathrm{xy}}|$, and $\mathcal{H}_\mathrm{Z}$ is a minor perturbation compared to the crystal field, only the low-$\langle J_\mathrm{z}\rangle$ states are substantially affected by the in-plane field. In the case of $\vec{B} = \vec{B}_{\mathrm{xy}}$, i.e., with no out-of plane component, the symmetry breaking is strongest and already a weak field will lead to $\langle J_\mathrm{z}\rangle \approx 0$ for all states of an integer J adatom in *any* given adsorption site.

It is reasonable to discuss the influence of the magnetic field on the time inversion properties only after the coupling to electrons has been introduced, since it greatly affects the dynamics as well which is due to scattering with these electrons.

3.2.2. Coupling to Electrons

An adatom is never completely isolated from (the substrate) electrons, especially not in the STM geometry where a conducting tip and substrate are needed. Often, the first effect of a band structure on the adatom is a change in its electronic configuration. Mostly the adatom will not have the same number of electrons as its free counterpart. Secondly, scattering with substrate electrons and, in the case of STM, also the tip electrons occurs. Of all the scattering processes, those involving an interaction between electron spin $\vec{s}$ and $\vec{J}$ are of interest here. It can be described [34] by the operator:

$$\mathcal{V} = \vec{s} \cdot \vec{J} = s_z J_z + \frac{1}{2} \left(s_+ J_- + s_- J_+ \right). \tag{3.6}$$

The resulting effect on the spectrum, again, cannot be easily stated. In order to quantify it, precise knowledge of the band structure would be needed, as well as the coupling strength between $\vec{s}$ and $\vec{J}$ which itself can depend on many parameters. Three qualitative statements can be made, however:

- The interaction between substrate electrons and the adatom's total angular momentum leads to a renormalization of the energy levels, although these shifts can rarely be observed directly [35].
- Scattering induces decoherence which, in turn, can have the effect of destroying the phase relations of the $\left|\Psi_M^{a/s}\right\rangle$. In this way, the splitting for integer J as in fig. 3.1 b) can be lifted and the steady states are again $|\Psi_{\pm M}\rangle$ (see section 4.5 and [36]).
- Scattering with electrons is the source of transitions between different $|\Psi_M\rangle$. This includes inelastic processes, such as excitations and relaxations via spin-flips, but also elastic magnetization reversal processes. These latter scattering events can lead to the formation of a Kondo-singlet and the complete screening of the adatom's magnetic moment if the temperature is low enough.

The details and consequences of the second issue will be discussed at a later stage. Instead, the following paragraphs will focus on the dynamics of the system when $\mathcal{H}_{\mathrm{CF}}$ determines the spectrum and scattering is considered as a weak, time-dependent perturbation.

The understanding of transitions between the different states of an adatom within the same multiplet relies on the analysis of the matrix elements

$$\langle \Psi_{M'} | \, J_{\mathrm{i}} \, | \Psi_M \rangle , \text{ where } i \in \{z, x, y\} \text{ or } \{z, +, -\}. \tag{3.7}$$

This procedure is applicable not only for interactions with electrons, as in eq. (3.6), but exchange-like coupling to any other angular-momentum type operator. This includes the lowest order interactions with nuclear spins, or direct exchange with other adatoms (if present), but in STM the dominant contribution comes from substrate and tip electrons.

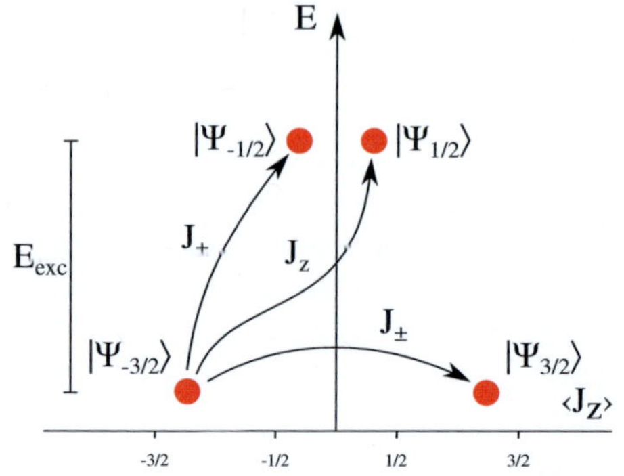

Figure 3.3.: *Transitions from one of the degenerate ground states to all other states in the system $J = {}^3\!/_2$ with $B_2^0 < 0$ and $B_2^2 \neq 0$. While only electrons with $E > E_{\text{exc}}$ can excite the system into one of the $\pm{}^1\!/_2$ states, all electrons, including the ones at low energy, can still lead to transitions between the two ground states and result in a zero net magnetic moment.*

Analyzing excitation processes in the low-temperature limit mainly deals with the tunneling electrons, and gives insights into which transitions should be observable with ITS. The correctly weighted sum of the three $|\langle\Psi_{M'}|\, J_{\text{i}}\, |\Psi_M\rangle|^2$ yields the transition probability for the processes $|\Psi_M\rangle \leftrightarrow |\Psi_{M'}\rangle$:

$$P_{\text{M,M}'} = \frac{1}{4}\left[|\langle\Psi_{M'}|\, J_+\, |\Psi_M\rangle|^2 + |\langle\Psi_{M'}|\, J_-\, |\Psi_M\rangle|^2\right] + \frac{1}{2}\,|\langle\Psi_{M'}|\, J_{\text{z}}\, |\Psi_M\rangle|^2 . \qquad (3.8)$$

This quantity allows to compare cross-sections for the transitions to different excited states. Likewise, it yields the number of possible excitations that could be found in a measurement or, inversely, may allow to estimate some crystal field parameters from the observed excitation spectra.

Closely related to excitations is the inverse process of relaxation. The above transition probabilities for relaxation from an excited state to different final states allow to predict possible paths for magnetization reversal. Furthermore, relaxation implies finite lifetimes of the excited states so that – due to the uncertainty relation – their energy is broadened which, in turn, has an influence on the line shape of the measured spectrum (see eq. (2.8)).

Coming back to the example of a $J = {}^3\!/_2$ system with an out-of plane easy axis and a nonzero in-plane anisotropy ($B_2^2 \neq 0$) as in fig. 3.2, calculation of the matrix elements reveals that not only the operators $J_\pm$ but also J_z contribute to inelastic processes, see fig. 3.3. Additionally, one finds that another process is possible with the $J_\pm$: elastic scattering between the ground states $|\Psi_{-3/2}\rangle$ and $|\Psi_{3/2}\rangle$. In other words, any electron at or near the Fermi level can induce transitions between the ground states. The occurrence of the transitions depends then on the electronic density of states, temperature, and coupling strength. At low temperatures, the consequence of this scattering process is the formation of a so-called Kondo-singlet, which manifests itself in the DOS as a narrow resonance at the Fermi level and is thus experimentally detectable in STS [4, 37]. Still, also at higher temperatures direct transitions between the ground states are allowed, only the Kondo-singlet will not be observable. Although having a Kondo system is interesting by itself, it is certainly incompatible with the quest to make atomic bits a reality and should be avoided from the viewpoint of this thesis.

3.3. The Role of Time Inversion Symmetry

Fortunately, there is a possibility to combine J and the adsorption site symmetry in such a way that one can have both — two degenerate non-coupling ground states and forbidden elastic spin scattering — and this scenario relies on the system's temporal symmetry to suppress the ground state transitions. In contrast to both previous examples, this needs a three-fold (or sixfold) rotationally symmetric site occupied by an integer-J adatom (with an appropriate value of J). Figure 3.4 shows two examples of this, with $J = 2$ and $J = 4$ in $C_{3\mathrm{v}}$. It is assumed that the $m = 0$ operators are such that an out-of plane easy axis is preserved, i.e., the ground states are the ones with maximum $\langle J_z \rangle$. Then the only further nonzero operator[4] is B_4^3, that partitions the $|\Psi\rangle$ states into three distinct groups (states in parentheses apply only for $J = 4$):

- $|\Psi^{\blacksquare}\rangle$ consisting of $|+1\rangle, |-2\rangle, (|+4\rangle)$
- $|\Psi^{\bullet}\rangle$ consisting of $|-1\rangle, |+2\rangle, (|-4\rangle)$
- $|\Psi^{\blacktriangle}\rangle$ consisting of $|0\rangle, (|\pm 3\rangle)$

Obviously, states from different groups are all pairwise orthogonal, independent of their position in the spectrum. The reason that the ground states are not connected by a single-electron scattering event also relies on time inversion symmetry, although neither system obeys Kramers' theorem. Rather, as a result of the subdivision induced by the operator B_4^3, the doublet states represented by red squares and blue circles with the same $|M|$ are related by the time inversion operator $\mathcal{T}$:

$$\mathcal{T}\,|\Psi_M^{\blacksquare}\rangle = \pm\,|\Psi_{-M}^{\bullet}\rangle\,. \tag{3.9}$$

To show this, consider the general properties of $\mathcal{T}$. First, $\mathcal{T}^2 = (-1)^{2J}$, which means $\mathcal{T}^2 = 1$ in an integer-J system. Secondly, it commutes with $\mathcal{H}_{\mathrm{CF}}$, so that $\mathcal{T}\,|\Psi_M\rangle$ remains an eigenstate of $\mathcal{H}_{\mathrm{CF}}$ with the same energy if $|\Psi_M\rangle$ is an eigenstate. With this, one can state that the action of $\mathcal{T}$ on one state is

$$\mathcal{T}\,|\Psi_M^{\blacksquare}\rangle = c_1\,|\Psi_M^{\blacksquare}\rangle + c_2\,|\Psi_{-M}^{\bullet}\rangle\,. \tag{3.10}$$

[4] Strictly, this is only true for $J = 2$, and for $J = 4$ if $\ell \leqslant 2$. For $\ell = 3$, the $O_6^{3/6}$ also appear, but they do not change the picture qualitatively since they couple only those states that were already affected in the same way by O_4^3.

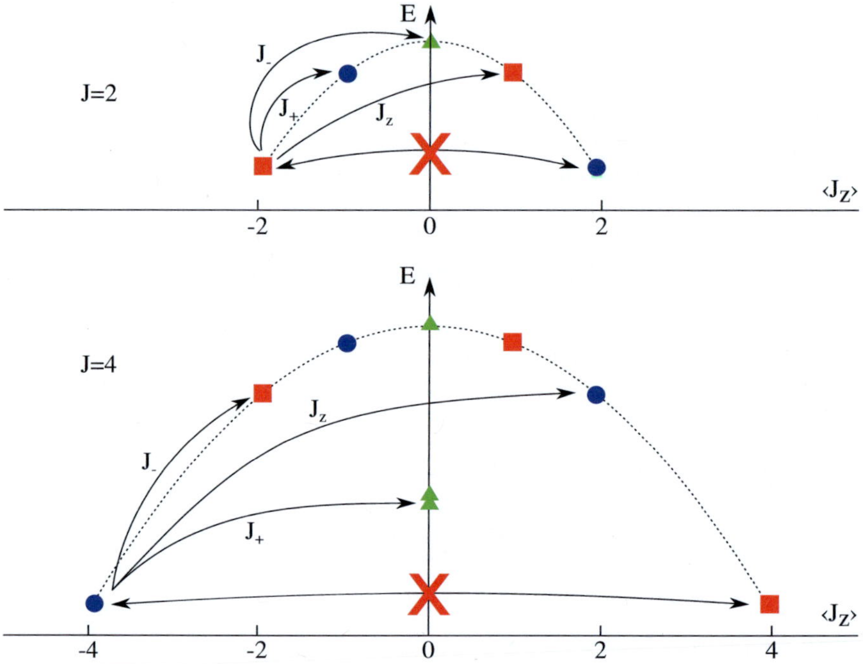

Figure 3.4.: *Spectra of an easy-axis $\mathcal{H}_{\mathrm{CF}}$ for $J = 2$ (upper panel) and $J = 4$ (lower panel) in a C_{3v} site. The colored symbols represent the three different groups of states. Some allowed excitation processes with the respective operator are shown, as well as the forbidden ground state transitions.*

When expressing $|\Psi_M^{\blacksquare}\rangle$ in the basis $|M\rangle$, one can use $\mathcal{T}\,|M\rangle = (-1)^M\,|-M\rangle$; a consequence of this is $c_1 = \langle\Psi_M^{\blacksquare}|\,\mathcal{T}\,|\Psi_M^{\blacksquare}\rangle = 0$ so that only the second term in eq. (3.10) remains. Using $\mathcal{T}^2 = 1$ fixes eq. (3.9) except for the prefactor $c_2 = \pm 1$.

Having explored the transformation of the states from the ▪ and • groups, one can use a few more identities: Both $\mathcal{T} J_{\mathrm{i}} = -J_{\mathrm{i}}\mathcal{T}$ for $i \in \{x, y, z\}$, and $\langle\chi|\,\phi\rangle = \langle\mathcal{T}\phi|\,\mathcal{T}\chi\rangle = \langle\mathcal{T}\chi|\,\mathcal{T}\phi\rangle^*$ can be used with eq. (3.9) to finally show that

$$\begin{aligned} \langle\Psi_{-M}|\,J_{\mathrm{i}}\,|\Psi_M\rangle = &\ \langle\Psi_{-M}|\,\mathcal{T}^2 J_{\mathrm{i}}\,|\Psi_M\rangle^* \\ = &-\langle\Psi_{-M}|\,\mathcal{T} J_{\mathrm{i}}\mathcal{T}\,|\Psi_M\rangle^* \\ = &-\langle\Psi_{+M}|\,J_{\mathrm{i}}\,|\Psi_{-M}\rangle^* \\ = &-\langle\Psi_{-M}|\,J_{\mathrm{i}}\,|\Psi_M\rangle = 0. \end{aligned} \tag{3.11}$$

Consequently, it follows that $\langle\Psi_{-M}|\,\mathcal{V}\,|\Psi_M\rangle = 0$ which, in turn, means that one-electron processes cannot lead to scattering from one ground state to the other!

Note that eq. (3.11) does not hold for half-integer systems. There, $\mathcal{T}^2 = -1$ and no such constraint for the matrix elements can be found. That is, the matrix elements may accidentally be zero, but there is no need for this to be true.

It is vital to note that all of the above is true only in the case of $\vec{B} = 0$. A non-vanishing magnetic field destroys the time inversion symmetry of the quantum system, and thus, eq. (3.9) which was used in eq. (3.11) does not hold any longer. Hence, also this derivation is no longer valid – in other words, transitions between the lowest states are gradually allowed with increasing $|B|$.

3.4. Validity of the Model, Viable Systems

As can be seen from the publication date of [31] (1953), the description of a magnetic impurity in a nonmagnetic host by a crystal field Hamiltonian, expressed in terms of Stevens operators is everything but new. Rather, it is a long established model that found its original application in magnetic resonance experiments. The importance of its rediscovery for single magnetic adatoms can be appreciated when one considers that most of the early work on the transition metal adatoms only used a lowest-order uniaxial spin Hamiltonian:

$$\mathcal{H} = DS_z^2 + E(S_+^2 + S_-^2) \qquad \text{or} \qquad \mathcal{H} = DS_z^2 + E(S_x^2 - S_y^2), \tag{3.12}$$

where D and E are the anisotropy constants[5] that are proportional to B_0^2 and B_2^2. This Hamiltonian does not include higher order operators (for transition metals up to $n = 4$ is possible), and therefore fails to capture the properties of an adsorption site with rotational symmetry other than C_{2v}. It is thus only appropriate in a very limited number of cases (see, e.g., fig. 3.2); in all other cases it can lead to false predictions and interpretations.

However, also the full theoretical description of a magnetic adatom with the above model is not universally applicable. The system's properties can be altered by several aspects which are not found in the model. As was mentioned before, the influence of the baths of electrons cannot be taken into account within this model. A master equation approach for this purpose will be described in section 4.4.2 and section 4.5.

More importantly, not all combinations of adatom and substrate allow to simply assume $L - S$ coupling. This is particularly true for the magnetically interesting $3d$ transition metals. Since their d-orbitals lie at the Fermi level and radially extend far outward, they are strongly affected by the crystal field, which acts on the spatial component of the wave functions. Thus, the orbitals hybridize, thereby reducing the orbital angular momentum L. In these cases, the intra-atomic spin-orbit coupling is not the dominant interaction and will only couple the remaining orbital angular momentum and the spin (see, e.g., [32]).

[5] either E is not identical in the two cases, or the definition of $S_\pm$ includes a factor of $1/\sqrt{2}$

Furthermore, the substrate may also acquire a polarization around the adatom. The magnitude of these effects can usually only be quantified with numerical methods. The crystal field Hamiltonian could still be used when J is defined to describe the total angular momentum of the whole impurity. However, it is not quite clear what ℓ should be, and the calculations often find values for L and S that, when combined, result in neither integer nor half-integer J values.

Both of these effects can be mitigated when turning away from transition metals towards rare-earth elements. There, the partially filled subshells are the 4f orbitals which are energetically lower than the Fermi level. With $\ell = 3$ and the possibility to combine the magnetic moments of a larger number of electrons than in d orbitals, higher magnetic moments of up to $\sim 10.5\,\mu_B$ can be expected. Furthermore, they are closer to the nucleus than the $5/6s$ and $5p$ orbitals and are thus well shielded from the conduction electrons. Finally, the nuclei are heavier compared to $3d$ elements which leads to stronger spin-orbit interaction. Thus, using rare-earth elements has the added benefit of similar or even higher magnetic anisotropy energy than in transition metals while the description with a CF-Hamiltonian is much more justified.

Another difficulty arises in transition metals when considering the actual number of relevant electrons in the system. As stated earlier, the oxidation state of an adatom can be different from a free atom and the determination might be a non-trivial task. Fortunately, most rare-earth atoms can be assumed to prefer their 3+ oxidation states, as they do in numerous compounds. This allows for a simple estimate of L, S, and J and subsequently the construction of $\mathcal{H}_{CF}$. Hence, while the crystal field method can in principle be used also for transition metal adatoms, it is – by comparison with rare-earth elements – difficult to identify the necessary properties and quantum numbers of the electronic shell. The most important advantage of transition metals is that their magnetic moments are easily accessible for inelastic tunneling experiments. In contrast, this is not necessarily true for $4f$ orbitals since their localized nature can render the acquisition of spectra difficult due to low signal intensity.

Chapter 4.

4 Holmium Adatoms on Pt(111)

After a summary of T. Schuh's experimental findings, the subsequent progress — experimental and theoretical — will be presented in detail. This includes further lifetime dependencies on voltage, and, most importantly, magnetic field. The supporting ab initio *calculations were only finished after T. Schuh's thesis and could then be analyzed. These results will shortly be discussed as well.*

Then the focus will shift towards further predictions and refinements of the model. Within the established framework, some remarks will be given on the influence of symmetry-breaking parameters, as well as current and temperature, RKKY- and hyperfine interactions. Furthermore, the viability of a classical and a full quantum master equation calculation will be discussed. Lastly, some open questions will be briefly touched on. Contributions by collaborating groups and individuals will be mentioned appropriately.

4.1. Summary of T. Schuh's Work

Motivation

After several results for STM experiments on transition metal adatoms [6, 38–40] had been published, Schuh et al. started to investigate rare-earth adatoms [10]. This shift away from transition metal elements was motivated by the possibility of having greater magnetic moments and well shielded magnetic subshells that should show less hybridization.

The first $4f$ element that was investigated was gadolinium (Gd) on platinum and copper (111) surfaces [41]. There, it was found that the adatom was in its $3+$ state, leaving the $4f$ orbitals with seven electrons which resulted in a ground state multiplet of $J = S = {}^{7}/_{2}$.

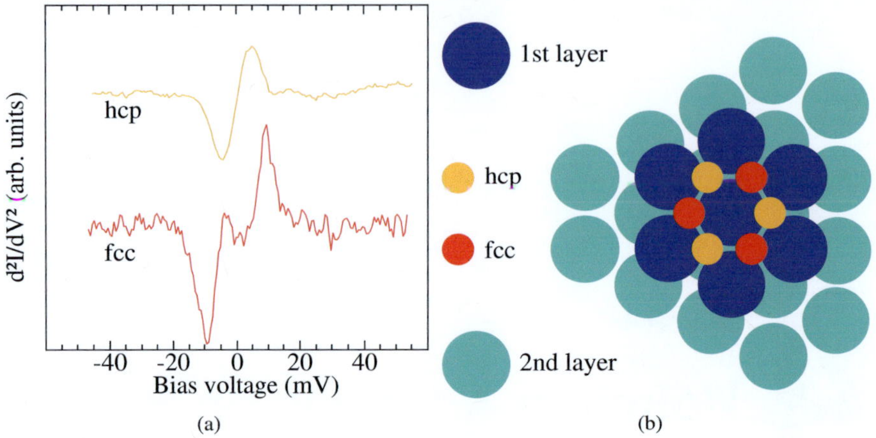

Figure 4.1.: *Background-corrected inelastic spectra on holmium atoms on Pt(111) in* fcc *and* hcp *adsorption sites, respectively (a). In the* hcp *case the dip-peak signature of an excitation overlaps at $V = 0$ so that an unambiguous determination of the excitation energy is not possible (after [10]). (b) schematically shows the* fcc *and* hcp *positions on a (111) surface.*

Similar to transition metal adatoms before, low-temperature inelastic spectroscopy measurements at $\sim 4\,\mathrm{K}$ revealed signatures of excitations at energies of a few meV which were attributed to spin excitations. Since the half-filled $4f$ subshell does not exhibit an orbital angular momentum, the magnetic anisotropy was explained with intra-atomic exchange interaction with the $5d$-shell.

The main results of the Gd experiments were that a) it is indeed possible to study the $4f$ electrons with STM, b) the observed magnetic anisotropy is comparable to that of transition metals, but c) with a reduced influence of the conduction electrons, regarding relaxation processes (longer lifetimes of the first excited state) and hybridization. Furthermore, the supporting *ab initio* calculations could be tested and verified. Lastly, although a $J = S = {}^7/_2$ system should be susceptible to Kondo exchange, this phenomenon was not observed which can be interpreted as a very low scattering probability of substrate electrons with the $4f$ orbitals, resulting in low T_K.

Experimental Results

From these findings, the idea to investigate holmium on a Pt(111) surface was inspired since in its 3+ state it exhibits the largest possible total angular momentum of all elements with $J = 8$ ($L = 6, S = 2$) which should further enhance the MAE. Additionally, assuming an out-of plane easy axis, the full crystal field Hamiltonian with all its O_n^m operators predicted that there should be two ground states that show neither coupling nor transitions between them via single-electron scattering.

The first characterization dealt with the adsorption sites and possible ways to distinguish between adatoms on *fcc* and *hcp* places. While their dI/dV curves were mostly identical in the bias range of ± 1 V, a discrepancy was found above $\sim$ 600 mV that allows to identify the adsorption site without having to determine the atomic positions. This is usually done by atomically resolved scanning across a step edge in order to deduce the position of atoms in the second layer from the surface. Since the necessary tip conditions might not always be attainable, a spectroscopic identification via a comparably simple STS measurement is certainly desirable.

Following this analysis, ITS revealed an excitation below 10 meV in both adsorption sites, with average values of 9.6 meV in the *fcc* site and 5.9 meV in *hcp* (see fig. 4.1). While in the *hcp* site the dip-peak pair could not be separated due to the experimental energy resolution, the spectra on *fcc* adatoms show that the width of the observed features was lower than the excitation energy so that there was no overlap at zero bias. This is another crucial requirement for obtaining a stable ground state because if the first excited state is broadened so much that it overlaps with the ground state, it can be excited thermally even at $k_B T \ll E_{\text{exc}}$. Consequently, from there on the experiments focused almost exclusively on *fcc* atoms. The excitation probability was found to be of the order of 5 % (for both sites), however, later a refined analysis of several d^2I/dV^2 spectra from the same data set obtained slightly lower values of only $\sim$2 %.

Since spin-polarized STM can detect magnetic contrast not only in extended structures but also different magnetization orientations in single atoms [42], the next series of experiments was devoted to the attempt to observe the two expected ground states of the system. It was hypothesized that Ho/Pt(111) — similar to the case of gadolinium — would be governed by an out-of plane easy axis; the expected spectrum is schematically shown in fig. 4.3(a).

For the experiments, an antiferromagnetic manganese-coated tungsten tip was used in order to perform spin-polarized measurements (see section 2.1.3, eq. (2.9) and (2.13)). Indeed, they revealed a contrast between two different states[1] as is shown in fig. 4.2(a). The detection was done via changes in the current or the lock-in signal, at a bias voltage of 5 mV and a modulation of 0.8 mV so that the voltage was always less than the excitation energy. As a result, the lock-in signal showed a 1 % difference between two constant values on an atom, but no contrast on the Pt substrate. Furthermore, it was found that the system could be reversibly switched between those two states by applying — with open feedback loop — a short voltage pulse at up to 65 mV for 200 µs, with a success rate of less than one.

Apart from the fact that two different atomic states could be observed which were interpreted as due to the magnetic ground states, further quantitative results could be obtained

[1] It is self-understood that much care was taken to relate this effect to a magnetic origin. No such contrast was observed with bare tungsten tips, neither on adatoms nor on the platinum substrate — this behavior could only be found when magnetically coated tips were used to probe the Ho atoms.

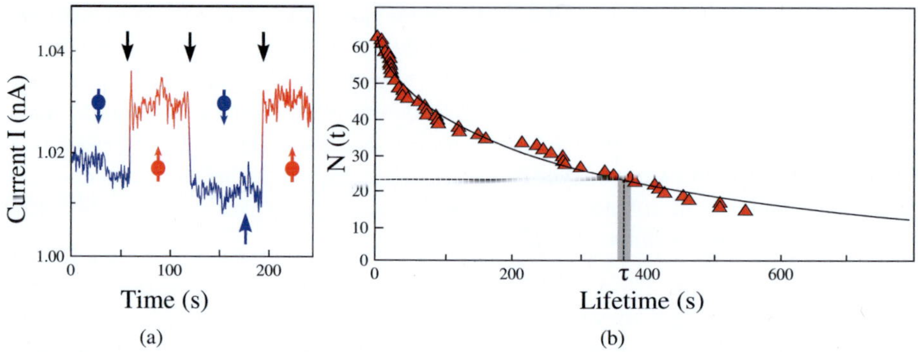

Figure 4.2.: *Switching behavior of an* fcc *adatom with voltage pulses (a). At the positions indicated by black arrows the system changed its state when a pulse was given; the blue arrow marks the time where a pulse did not lead to switching. In (b) the lifetimes of states after an induced switch is plotted as a decay of a population. An exponential fit yielded a mean lifetime of* $\tau = 373 \pm 10$s. *(Adapted from [10]).*

Apart from the fact that two different atomic states could be observed which were interpreted as due to the magnetic ground states, further quantitative results could be obtained from the pulse experiments. First, the switching probability was found to increase to nonzero values just around a threshold of ~10 mV — very close to the observed excitation energy, as is described in more detail in section 4.4.2. This is a strong indication that the excited state plays an important role in the switching process. Further, the switching probability did not reach 50 % but a lower value which was then attributed to the influence of the spin-polarization of the tip. The suggested path for switching between the two ground states was via the $|\Psi_{\pm 7}\rangle$ states, with a small but nonzero probability to tunnel through the anisotropy barrier. Two different one-electron scattering processes are needed to switch from one ground state to the other (see fig. 4.3(b)): one "spin-conserving" with J_z and one "spin-changing" with $J_{\pm}$. For example, the latter process can excite the system from $|\Psi_{-8}\rangle$ to $|\Psi_{-7}\rangle$ with J_+, which will then either relax back with J_- or tunnel to $|\Psi_8\rangle$ with the help of J_z. Of course, also the opposite sequence is possible, i.e., an excitation with J_z through the barrier and a subsequent relaxation with J_+. If higher energies are available also the second set of excited states $\left|\Psi_6^{a/s}\right\rangle$ can lead to magnetization reversal: The relaxation from there to either side is equivalent and thus this path should have a 50 % probability to end in $|\Psi_8\rangle$ or $|\Psi_{-8}\rangle$.

The other interesting quantity that could be extracted from the pulse measurements is the lifetime of the two different conductance states. It was found that some time after the pulse the system spontaneously switched (back) to the initial state. The corresponding times ranged from several seconds up to a few minutes. In some cases the lifetime even outlasted the available period to record a time trace of ~8 min. From the lifetime data a population-like exponential decay (see fig. 4.2(b)) was fitted and gave a mean lifetime of ~370 s. In comparison to any previously measured spin lifetime of magnetic adatoms,

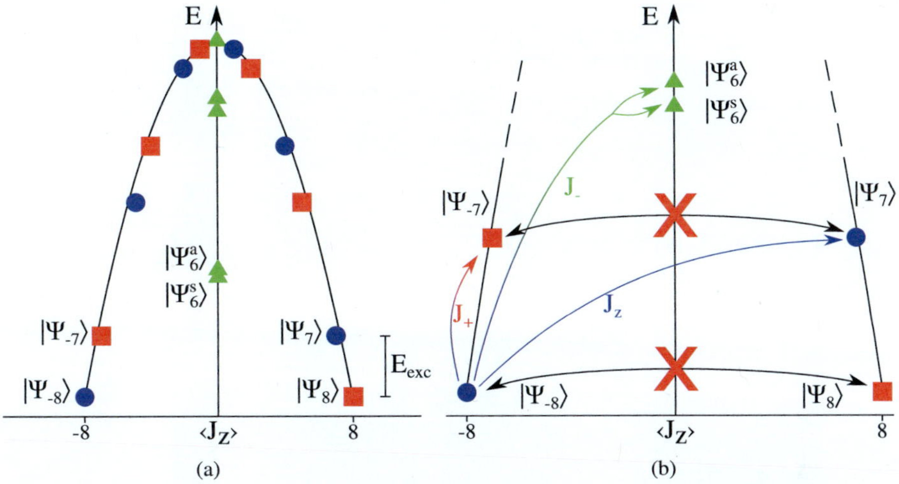

Figure 4.3.: *Schematic of the $\mathcal{H}_{CF}$ spectrum (a) and some of the allowed transitions between the lowest six states (b). Similarly to the examples $J = 2, 4$ the two ground states belong to different groups of states (■ and ●), but this is also true for the doublet of the first excited states. It is only the next pair of states that forms (anti)symmetric, energy-split states (▲). The transitions in (b) are labeled with the operator $J_i, i \in \{z, +, -\}$ that has the dominant/only contribution to this process. Again, no operator can lead to transitions between the ground states.*

this means an increase in lifetime between seven and nine orders of magnitude! Although this result should initially be taken with the appropriate skepticism, the model gives an explanation for it: Holmium atoms on Pt(111) occupy three-fold hollow sites which, in turn, establish a C_{3v} symmetry. As was discussed for the examples in section 3.1.3, this system also exhibits the advantageous properties of non-coupling ground states and forbidden ground state transitions via single-electron scattering which is illustrated in fig. 4.3. Thus, all magnetization reversal processes *must* involve excited states and are therefore exponentially suppressed at low temperatures.

4.2. New Experimental Results

4.2.1. Modification of the Measurements

After the initial findings of Schuh et al. and their preliminary interpretation, more experimental work was needed in order to substantiate the claims. To this end, an investigation of the lifetime dependence on bias voltage and magnetic field was conducted, which yielded also some information about the influence of tunneling current and temperature.

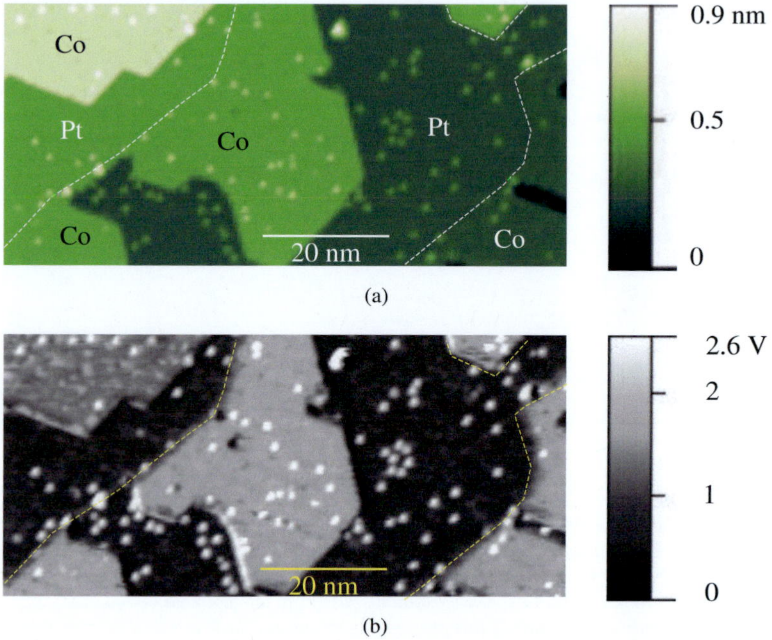

Figure 4.4.: *Detail of the topography (a) and spin-polarized dI/dV map (b) of the Co/Pt(111) system, decorated with Ho atoms that appear as small bright dots. The Pt step edges are marked with dashed lines. (b) shows a contrast between the Co region in the upper left corner and all other Co islands, while both show a signal that differs from the Pt substrate (*300 mV, 1.5 nA, $V_{\mathrm{mod}}^{\mathrm{rms}}$ =8 mV *at* 5.3 kHz*).*

Although the initial experiments with a Mn-coated tungsten tip yielded some usable results, a slight modification in the measurement procedure was made. One of the difficulties during the measurements was to maintain the structural integrity of a tip that was sufficiently spin-polarized to show the 1 % conductance change during switching. Thus, chromium was chosen as a coating material since greater mechanical stability was expected. Additionally, small amounts of cobalt (less than a monolayer) were deposited to the platinum substrate slightly above room temperature before the holmium deposition. The cobalt islands that form during this procedure are known to be magnetized out-of plane [42] and can thus be used as an independent reference to confirm a suitable spin-polarization of the tip. This is illustrated in fig. 4.4. However, the additional cobalt islands have a magnetic stray field which has a detrimental effect on the lifetimes, and an indirect influence on the Ho adatoms via RKKY interaction might occur [42]. Therefore, this procedure was only used occasionally, mostly with newly prepared tips, to assign the observed dI/dV contrast of Ho atoms to a magnetic origin.

Having confirmed that the previously found switching could also be seen with Cr tips, the various dependencies were studied. First, the results of the magnetic field dependence on the lifetime will be presented. After that the effect of bias voltage on the lifetimes will be discussed. It shall be noted that the excitation energy found in renewed ITS experiments yielded slightly lower values of E_{exc} ~8 meV instead of 9.6 mV for *fcc* atoms. As the older measurements were taken with a different STM, there might be a small discrepancy in the bias calibration, one possible deviation coming from the use of different voltage dividers. Apart from that, the older measurements on several atoms gave varying results, ranging from ~7 meV to 11 meV, which is attributed to a slightly different local environment, while the newer ones showed a stronger concentration of excitation energies and the mean value of all the fits amounts to 7.68 ± 0.01 meV. One example is given in fig. 4.8.

The determination of the lifetime of the first excited state was done as is described in appendix A for all measured atoms and spectra which resulted in an average value of

$$\tau_{|7| \rightarrow |8|} = (470 \pm 30)\text{fs}. \tag{4.1}$$

The statistical error seems small, but it does not reflect that extracting the lifetime from a spectrum was applicable only for a subset of the measured ITS curves. For some spectra, the calculated intrinsic width would approach zero and thus give extremely long lifetimes; in few cases, the calculated intrinsic width would even become negative, making the interpretation meaningless. The reasons in both cases are not yet understood, and may only be found with a renewed measurement at better energy resolution. Conversely, the fits with narrow linewidth could represent the actual situation and the measurements that give the above lifetime value could be broadened due to additional noise (e.g., from electronics). Therefore, the above mean value can only be seen as a lower limit. As a last remark, the excitation cross section in the new ITS experiments was roughly 1 %.

4.2.2. Magnetic Field Dependence

Experiment

The most important experiment was to measure the lifetime dependence of the ground states on an external magnetic field. As the whole model and the tentative interpretation rely on the assumption that the observed excitations and the spin-polarized contrast are of magnetic origin, detecting a marked dependence of the ground state lifetime on B is a necessity. All other possible excitations (e.g., phonons), or switching phenomena (e.g., structural changes of the tip, hopping between different adsorption sites) are not expected to exhibit any impact of (weak) magnetic fields.

For the investigation of the field dependence, a sample without additional Co was chosen. All measurements were conducted on a relatively isolated atom that was at least ~6 nm away from any other adsorbate or surface structure. Again, the bias used for detection

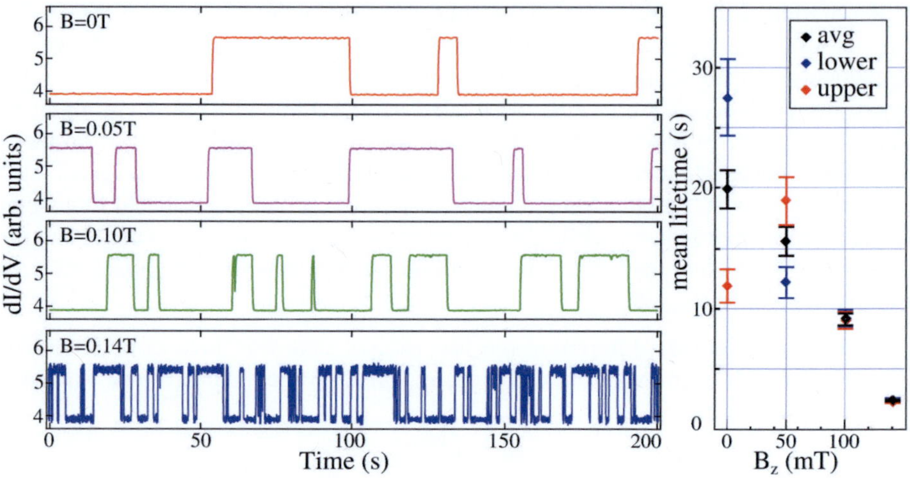

Figure 4.5.: *Influence of magnetic field on the lifetimes of the ground states. The left hand side shows typical time traces of the spin-polarized signal at four different fields. Clearly, with higher fields the switching frequency increases. Although it cannot be stated with certainty which of the two observed dI/dV-values corresponds to the (anti-)parallel configuration of spin and tip polarization, at least the upper and lower conductance value can be analyzed separately since the tip did not change during the measurement. The quantitative result is shown on the right: the overall trend of shorter lifetimes with increasing field is confirmed.*

(5 mV) was low enough to mostly avoid excitations by the tunneling electrons; even when adding to that the modulation voltage of 0.8 mV (rms), the resulting maximum voltage was still below E_{exc}/e so that switching via the first excited state had a low probability.

A low field was then applied along the z-direction, which could only range from 0 to 140 mT due to experimental limitations at that time. Unlike in the earlier experiments by Schuh, this time the lifetimes were measured by observing the spontaneous switching between two states of the adatom[2], represented by different conductance values in the dI/dV signal, instead of using voltage pulses to induce a switch in the first place. Furthermore, the tunneling current in the new experiments was 50 nA — significantly higher than before — in order to improve the signal-to-noise ratio. A much clearer contrast of about 40 % in the switching behavior could be obtained, compared to only 1 % in the original experiments. Lastly, the temperature during the measurements was roughly 700 mK instead of $\gtrsim$4 K, which should have reduced the number of thermally activated processes.

The results of this measurement are displayed in fig. 4.5. While the four panels on the left illustrate a clear trend towards shorter lifetimes with increasing field, the quantitative results for the mean lifetimes are presented in the right panel. The experiment revealed

[2] A similar experiment with Fe clusters of five and six adatoms was conducted by Khajetoorians et al. [43] that also showed spontaneous switching with lifetimes ranging from milliseconds to a few hundred seconds.

decreasing average lifetimes (shown in black) going from ~20 s at zero field to ~2 s at maximum field. For the weakest fields, a deviation from this behavior was found: When going from 0 T to 50 mT the analysis revealed an increase of the lifetime of the state that corresponds to the higher conductance value. Also, the lifetimes of the two states are clearly different at zero field, but, admittedly, the number of data points is too low to make a general statement about the lifetime behavior at low fields[3].

Due to different experimental conditions with regard to the work of Schuh (section 4.1) the lifetimes cannot be compared directly. It is assumed that the much smaller[4] values found here result from the strongly increased tunneling current, with possible minor contributions from the changed tip material.

Analysis and Explanation

Qualitatively, the experimentally observed reduction of the lifetime with increasing field can be explained from the theoretical model, as was already discussed in section 3.3: The long lifetimes are a consequence of the protection by time inversion symmetry; when $\vec{B} = 0$, transitions between the ground states via single-electron scattering are forbidden. A nonzero field breaks time inversion symmetry and gradually allows these processes with increasing field. In other words, finite ground state[5] lifetimes are expected that decrease with higher fields. A brief quantitative analysis of this behavior will be discussed in section 4.4.1.

Of course, the lifetimes are not infinite even when the experimentally controllable magnetic field is set to zero. This is because a) the assumption of $\vec{B} = 0$ is never completely fulfilled, b) there are always a few thermally induced transitions to the first excited state, and c) the magnetic tip itself may have an influence on the atom. Although at low temperatures the magnetization reversal via the excited states is energetically suppressed for both tip and substrate electrons, one can argue that the tunneling electrons have a significant influence on the lifetimes, and furthermore that asymmetries of the two ground state lifetimes should be expected. Both will be discussed in the following paragraphs.

During the measurement, the adatom and its total angular momentum are coupled to electron reservoirs: On the one hand, "high-energy" ($eV \gg k_{\mathrm{B}}T$) electrons tunnel from the tip to the surface. On the other hand there are low-energetic substrate electrons [6] in thermal equilibrium. Since in this case the DOS of substrate electrons is high one might wonder whether the tunneling electrons can be neglected in the process of magnetization reversal. From the following qualitative argument, it can be understood that this is

[3]The crossing of the upper and lower state lifetimes at low fields is not statistically significant.

[4]From the viewpoint of statistical analysis shorter lifetimes can be considered beneficial, since this resulted in higher number of switching events (between ~70 and ~300).

[5]The terms "ground states" and "ground state transitions" will be used although only the lower of the $|\Psi_{\pm 8}\rangle$ is truly the ground state. This one will be denoted GS, the other GS*.

[6]Mathematical descriptions will be presented in section 4.4.2 and section 4.5.

not necessarily accurate when $\vec{B} \neq 0$: At low temperatures, the available phase space volume for inelastic scattering of substrate electrons (for both excitation and relaxation processes) is limited to a narrow shell at the Fermi surface. The energy scale for the "thickness" of that shell is given by the Zeeman splitting ΔE_Z between the two ground states. In contrast, the tunneling electrons, which have an energy far greater than the Zeeman splitting or the thermal energy, can scatter into all empty states above the Fermi level up to their initial energy. Therefore, when a magnetic field is applied in principle all tunneling electrons — at least those with the proper spin orientation — could lead to a ground state transition. Thus, although the occurrence of scattering with substrate electrons may be much higher than for tunneling electrons, the latter are expected to have a higher scattering cross-section. It should therefore come as no surprise that the tunneling current has a significant influence on the lifetimes.

Regarding an asymmetry of the lifetimes, again the contributions from substrate and tunneling electrons must be combined. Substrate electrons will preferably lead to relaxation from GS* to GS, and only rarely to the excitation in the opposite direction. Thus, the population ratio of both states in thermal equilibrium should be given by a Boltzmann factor $e^{\Delta E_Z/k_B T}$. The tunneling electrons, however, have an almost equal probability for the transitions in both directions — at least from an energetic point of view — effectively weakening the asymmetry. But, in order to obtain a contrast in the measurement, a spin-polarized tip is needed. Hence there is an imbalance between the spin orientations of the tunneling electrons leading to preferred ground state transitions in one direction. This effect is known as spin-transfer torque (STT) and is independent of B (antiferromagnetic tips were used). Thus, as long as the tunneling current has a non-negligible influence on the lifetimes, an asymmetry has to be expected, but its magnitude is a combination of the STT effect and the thermal equilibrium asymmetry.

Lastly, another influence of the measurement on the adatom and thereby the ground state lifetimes must be seen in the tip itself. First, its presence should alter the crystal field parameters. Although a quantitative account cannot easily be given, this will almost certainly disturb the symmetry since a perfect positioning cannot be achieved. How this may affect the lifetimes will be explored in some detail in section 4.4.1, but qualitatively it can already be expected that this will allow transitions between the ground states by scattering and also lead to a reduction of their lifetimes. Secondly, the tip itself is also a magnetic object. Despite the fact that antiferromagnetic tips are chosen, a minute stray field remains and may become important for the small tip-sample distances necessary in STM.

4.2.3. Voltage Dependence

In a second series of measurements, the voltage dependence of the lifetimes was studied. The reasoning behind it was that the tunneling electrons are the ones with the highest available energy in the system. Thus, the voltage should have an effect on the lifetimes

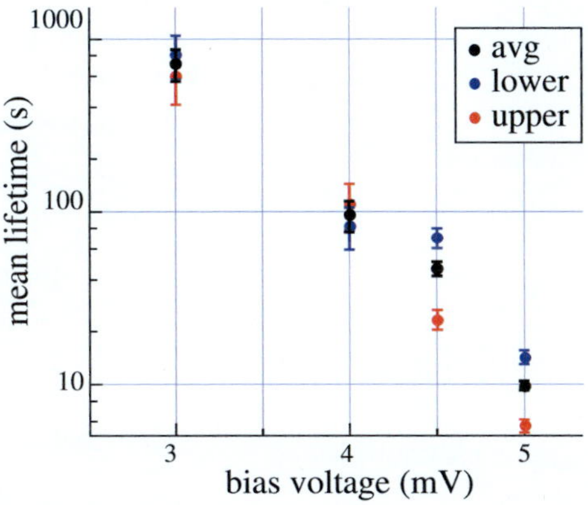

Figure 4.6.: *Bias dependence of ground state lifetimes. A dramatic decrease of lifetimes is observed when higher bias voltages are applied, i.e., when V is chosen closer to the excitation energy E_{exc}/e.*

when approaching the excitation energy, as thermally activated switching is most likely to come from the tunneling electrons instead of the substrate electrons[7]. The experiments were conducted at 700 mK, this time again at a low current of 1 nA, and with an added modulation voltage of 0.8 mV (rms). No magnetic field was applied, but in this case, Co islands were present on the substrate. Their stray field supposedly reduces the lifetimes in just the same way as an external magnetic field which is why the observed lifetimes do not match those of the previous experiments.

The results of the bias dependence measurements are presented in fig. 4.6. Again, a strong dependence can be seen, following roughly a reduction by almost one order of magnitude for a 1 mV increase in bias. Although qualitatively such an exponential behavior is expected for any kind of thermally activated process, a deviation from it would certainly cast doubts on the suggested magnetic origin of the observed contrast.

Once more, a small discrepancy of the lifetimes of the upper and lower conductance values is apparent for 4.5 mV and 5 mV. One possible explanation could involve the Zeeman splitting due the stray field of the Co islands that were less than 5 nm from the atom. Assuming a field parallel to the surface normal and estimating its magnitude for thermal equilibrium yields $B_{\mathrm{Co}} \approx 50\,\mathrm{mT}$. Admittedly, this should then also apply for the lower bias measurements. For 3 mV and 4 mV, however, the number of recorded switching events is only about one tenth of the higher bias situation (ca. 10 to 15 compared to ca. 100 to 140). Thus, the effect might not be seen due to poor statistics. In the framework of the theoretical model the aforementioned spin-transfer torque effect cannot explain the observed asymmetry since the fraction of electrons with sufficient energy to excite the atom to one of the $|\Psi_{\pm 7}\rangle$ states is too small by several orders of magnitude. Only if one substitutes the measured temperature with a much increased effective

[7] bias polarity was chosen such that electrons tunnel from tip to substrate

temperature, the STT effect could possibly describe the observed difference in lifetimes. However, unrealistically high values for T_{eff} would be needed. A proper treatment with a quantum master equation approach (see section 4.5) manages to capture this behavior with less severe assumptions.

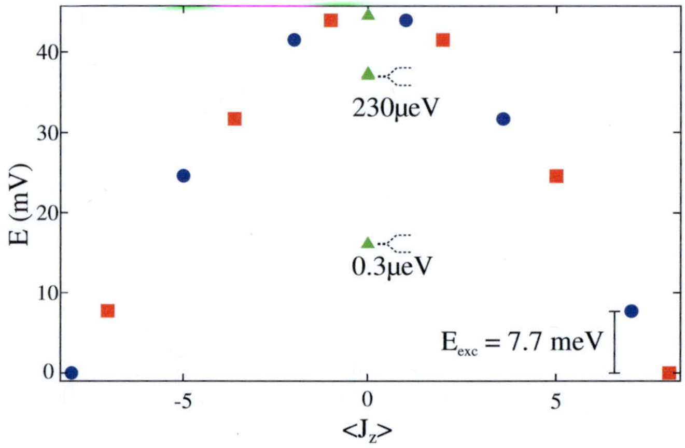

Anisotropy constant	value
B_2^0	-239 µeV
B_4^0	86 neV
B_4^3	293 neV
B_6^0	0.187 neV
B_6^3	-1.968 neV
B_6^6	0.630 neV

Figure 4.7.: *Spectrum of $\mathcal{H}_{\text{CF}}$ with the crystal field parameters from first principles (table on the right). As before, the three distinct groups of states are indicated with colored markers. The minute splittings of the $|\Psi_{3/6}^{\blacktriangle}\rangle$ due to the mixing operators are given in the diagram, as well as the splitting between ground state and first excited state.*

4.3. First Principles Calculations

Already the gadolinium experiments of T. Schuh were supported by *ab initio* calculations, performed by A. Ernst et al. at the Max-Planck Institute of Microstructure Physics in Halle (Saale) — a collaboration that was continued for the investigation of holmium. The calculations combined several different computational methods to determine the structural, as well as the electronic and magnetic properties of the Ho impurity. A noteworthy feature of the calculations is the possibility to account for strongly localized $4f$ states (see supplementary of [9]). The main results, though not the focus of this thesis, will be briefly summarized since they were essential for the corroboration of the tentative interpretation and further predictions.

First of all, the electronic configuration of Ho adatoms on Pt(111) was confirmed to be [Xe] $4f^{10}$, both for *fcc* and *hcp* adsorption sites. That means, the atoms give away their two $6s$ electrons and one from the f subshell. Based on this information and Hund's rules, the ground state multiplet is expected to have $J = 8$, resulting from $L = 6$ and

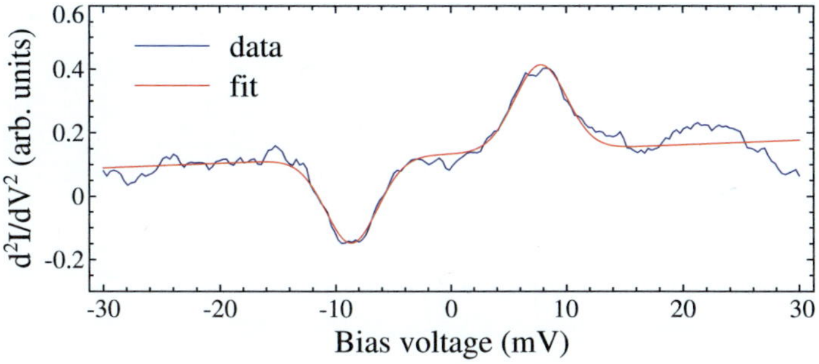

Figure 4.8.: *New ITS curve and corresponding fit of an* fcc *Ho adatom. The fit yields an excitation energy of* $8.0 \pm 0.1\,\mathrm{meV}$. *Aside from this dip-peak pair, no other inelastic signatures are visible. Experimental parameters:* $T = 4.4\,\mathrm{K}$, $I = 30\,\mathrm{nA}$ *at* $V = 30\,\mathrm{mV}$, $V_{\mathrm{mod}} = 2.4\,\mathrm{mV}$ *at* $16.4\,\mathrm{kHz}$.

$S = 2$, which also gives a Landé g-factor of $g_8 = 1.25$ and a total magnetic moment of $10\mu_\mathrm{B}$.

Second, the magnetic moments were extracted from the calculations which yielded $4.1\mu_\mathrm{B}$ ($3.9\mu_\mathrm{B}$) and $5.6\mu_\mathrm{B}$ ($5.45\mu_\mathrm{B}$) for spin and orbital contribution, respectively, in the *fcc* (*hcp*) adsorption site. Thus the calculation gives a good agreement with Hund's rules and justifies their preliminary use. Furthermore, the B_n^m parameters could be calculated, from which the model Hamiltonian $\mathcal{H}_\mathrm{CF}$ could then be set up and analyzed. The values of the anisotropy constants are presented in fig. 4.7 together with the resulting spectrum of $\mathcal{H}_\mathrm{CF}$.

With the anisotropy constants from the *ab initio* calculations available, the crystal field Hamiltonian could be tested against the experimental data. The easiest comparison of experiment and calculation considers the spectrum of $\mathcal{H}_\mathrm{CF}$ and the predicted excitation energy for the transition from $|\Psi_{\pm 8}\rangle$ to $|\Psi_{\pm 7}\rangle$. The calculated value of 7.7 meV almost perfectly matches the results from newer ITS experiments (see fig. 4.8) and gives still good agreement with the older values by Schuh et al. (fig. 4.1(a)). Since the spectrum has an overall parabolic shape and the calculated ground states are the $|\Psi_{\pm 8}\rangle$ states, further excitations might appear in inelastic spectra at higher energies. However, the calculation of the excitation cross section yields that reaching the second excited state has a very low probability compared to the first (a reduction of $\sim 10^{-5}$). Since the first excited state led to a conductance change of only $\sim$1 %, ITS cannot be expected to reveal the next excitation at 16.3 meV. In fact, no corresponding second dip-peak pair was observed, although admittedly the experimental resolution was not good enough to exclude two overlapping excitations if they were to appear at roughly the same energy.

Lastly, the electronic density of states, shown in fig. 4.9, reveals that all $4f$ states are $\gtrsim$2 eV away from the Fermi level and thus multivalent behavior can safely be ruled out. Moreover, it is found in the calculations that the Ho density of states is spin-polarized at the Fermi level (see inset of fig. 4.9), which is a crucial requirement for the interpretation

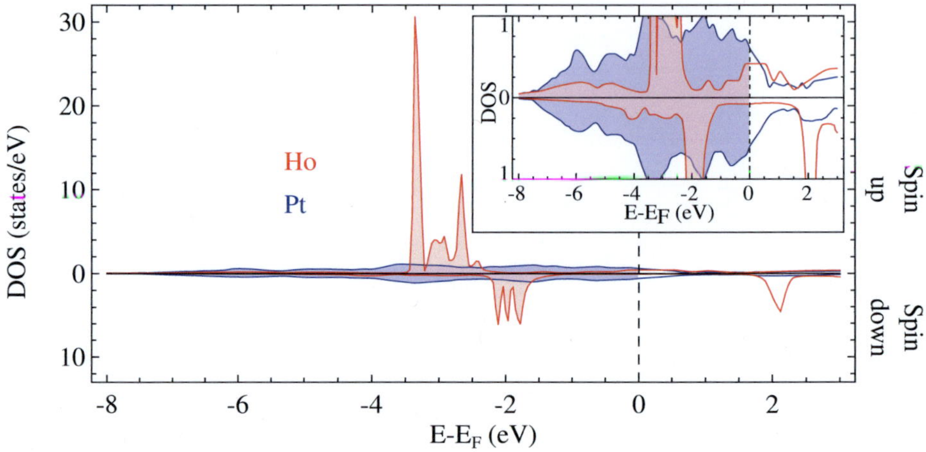

Figure 4.9.: *Calculated electronic DOS of Ho adatom (red) and Pt substrate (blue) showing both spin channels ("up" and "down"). The inset shows the low-DOS range in more detail, revealing that the Ho states are spin-polarized at the Fermi level ($\eta \approx 0.7$), whereas the Pt DOS is largely unpolarized.*

that the observed contrast in the spin-polarized measurements originates from switching between the two ground states $|\Psi_{\pm 8}\rangle$.

4.4. Analysis of Model Predictions

Since no further excitations are expected to appear in inelastic spectroscopy, the comparison with the calculated anisotropy constants is limited to evaluating the accuracy of B_2^0. But with the calculated values of all other B_n^m the dynamics of the system can be studied quantitatively. To this end, a classical master equation description was set up in order to calculate transition rates between states and their lifetimes, which will take up the greater part of this section.

Before that, it will shortly be discussed whether the *ab initio* B_n^m can actually be used to confirm the previously predicted influence of symmetry-breaking terms. The effect of additional low-*n* operators that perturb the spatial symmetry will be considered. Breaking of time-inversion symmetry is evaluated by calculating the magnetic-field dependence of transition matrix elements.

4.4.1. Influence of Symmetry Breaking

Owing to the threefold rotational symmetry the system exhibits two protected ground states. If the adatom's environment is disturbed, e.g., by lattice distortions, nearby ad-

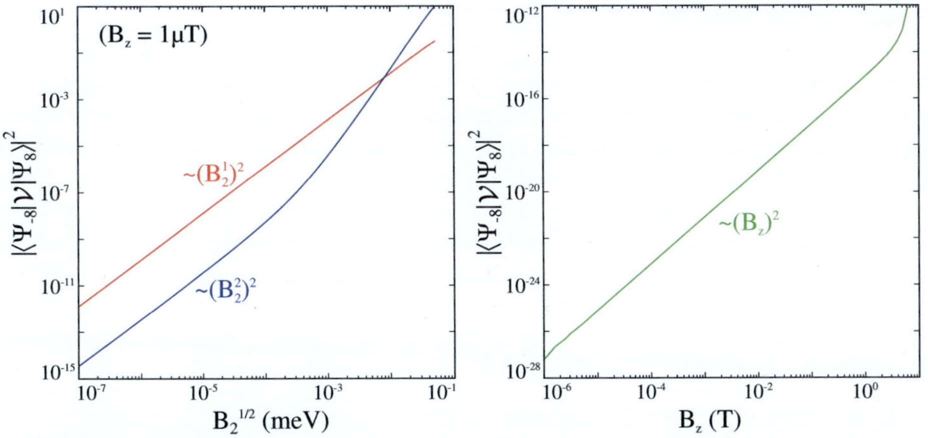

(a) (b)

Figure 4.10.: *Calculated influence of the operators O_2^1 and O_2^1 (a) and a B_z-field (b) on the transition matrix element of $\mathcal{V}$ between the two ground states. The symmetry-breaking is parametrized by the anisotropy constants B_2^1 and B_2^2, respectively. In the regime of weak perturbation $|\langle\Psi_{-8}|\mathcal{V}|\Psi_8\rangle|^2$ grows quadratically with B_2^m; above ca. 10^{-4} meV B_2^2 leads to an even stronger increase. For numerical stability, a minute magnetic field in z-direction of 1 μT was included in the calculation. The effect of an external magnetic field (b) is largely as expected — $|\langle\Psi_{-8}|\mathcal{V}|\Psi_8\rangle|^2$ grows quadratically until the Zeeman splitting of the ground states approaches E_{exc}.*

sorbates, or the tip, the symmetry is reduced and additional terms will appear in $\mathcal{H}_{\mathrm{CF}}$. Since the symmetry-breaking of the threefold hollow site reduces the system in general to a C_1 situation, only rotations by 360° lead to an equivalent description. Consequently, *all* operators up to sixth order would have to be considered in $\mathcal{H}_{\mathrm{CF}}$. However, since the anisotropy parameters can hardly be estimated and the operator O_2^1 already captures the mixing of all states, the influence of varying B_2^1 on the transition matrix element is studied (see fig. 4.10(a)). For comparison the effect of a nonzero B_2^2 is also shown — in a realistic scenario, both anisotropy parameters (among all others) will be small but nonzero. Unsurprisingly, both operators lead to a quadratic increase of $|\langle\Psi_{-8}|\mathcal{V}|\Psi_8\rangle|^2$ for weak B_2^m. If $B_2^2 \gtrsim 10^{-4}$ meV, its effect becomes even more pronounced.

A quantitative study of the influence of a B_z-field is shown in fig. 4.10(b). As qualitatively predicted from symmetry considerations, the formerly vanishing matrix element gradually grows with increasing field. The dependence of the matrix element is linear (i.e., quadratic in the transition rates) for weak fields, but changes drastically once the $|\Psi_{\pm 7}\rangle$ aligns with the former ground state $|\Psi_{\mp 8}\rangle$ from the other side of the barrier.

In conclusion, it can safely be stated that the calculated anisotropy parameters are compatible with the qualitatively expected behavior. It is thus reasonable to further investigate the sytem's dynamics with this set of parameters.

4.4.2. Dynamics and Rate Equations

The remainder of this section deals with the attempt to quantitatively test the anisotropy parameters against the experimental lifetime data. This was done with a classical rate equation approach. An approach like this was used to describe spin excitation spectra of magnetic adatoms by Hirjibehedin et al. in 2007 [38], and later, more elaborately, in 2010 by Loth et al. [44]. It was also used by Hübner et al. in 2014 [45] to study the dynamics of three adatom or cluster systems in different adsorption symmetries, among them the case of Ho/Pt(111) presented here. A critical view of the rate equation approach will be given, concluding that a full quantum master equation treatment is needed. This analysis was done in a collaboration with theory colleagues from KIT (Karlewski et al. [36]). Some details of the newer approach will be summarized in section 4.5.

In the rate equation approach, the time evolution of the population N_i of a state $|\phi_i\rangle$ is described by all transitions to final states $|\phi_j\rangle$, and can be separated into three contributions:

$$\frac{dN_i}{dt} = \sum_j (\Gamma_{ji}N_j - \Gamma_{ij}N_i) \tag{4.2a}$$

$$\sum_{j\neq i}\Gamma_{ij} = \sum_{j\neq i}\left(\Gamma_{ij}^{T\to S} + \Gamma_{ij}^{S\to T} + \Gamma_{ij}^{S\to S}\right) = \tau_i^{-1}, \tag{4.2b}$$

where the upper indices indicate the origin and destination of the scattering electron (T: "tip", S: "substrate")[8]. The $i = j$ terms are allowed and account for elastic scattering, without the atom's state changing. The lifetime τ_i of the initial state is then the inverse of the sum of transition rates Γ_{ij} to all other states.

In order to evaluate the transitions induced by scattering with electrons, the matrix elements of the operator $\mathcal{W} = \vec{J}\cdot\vec{s} + u$ will be considered, where $\vec{s}$ denotes the single electron spin and u an elastic contribution (compare eq. (3.2.2)). Remembering the spin-polarization of the tip η from eq. (2.10) and letting $\sigma = \pm 1/2$ denote the z-component of the scattering electron's spin, one can define

$$P_{ij}^{S\to S} = \sum_{\sigma\sigma'}\left|\langle\Phi_j\sigma'|\,\mathcal{W}\,|\Phi_i\sigma\rangle\right|^2 \tag{4.3a}$$

$$P_{ij}^{S\to T} = \sum_{\sigma\sigma'}\left|\langle\Phi_j\sigma'|\,\mathcal{W}\,|\Phi_i\sigma\rangle\right|^2\left(\frac{1}{2} + \eta\sigma'\right) \tag{4.3b}$$

$$P_{ij}^{T\to S} = \sum_{\sigma\sigma'}\left|\langle\Phi_j\sigma'|\,\mathcal{W}\,|\Phi_i\sigma\rangle\right|^2\left(\frac{1}{2} + \eta\sigma\right). \tag{4.3c}$$

[8] An interaction of tip electrons that re-enter the tip will be neglected since it is expected to be much weaker than the $S \to S$ term. The reason for this is that such an electron would have to cross the tunneling barrier twice, one time assisted by the bias voltage but the other time against it.

With this, as well as the effect of temperature on the occupation of states, and two parameters describing the coupling strengths to electrons within the substrate c_{ss} and tunneling electrons c_{ts}, the scattering rates can be related to $\mathcal{W}$ in the following way:

$$\Gamma_{ij}^{S\to S} = c_{ss}P_{ij}^{S\to S}\int_{-\infty}^{\infty} dEf(E)(1-f(E+E_i-E_j)) \tag{4.4a}$$

$$\Gamma_{ij}^{S\to T} = c_{ts}P_{ij}^{S\to T}\int_{-\infty}^{\infty} dEf(E)(1-f(E+E_i-E_j-eV)) \tag{4.4b}$$

$$\Gamma_{ij}^{T\to S} = c_{ts}P_{ij}^{T\to S}\int_{-\infty}^{\infty} dEf(E)(1-f(E+E_i-E_j+eV)), \tag{4.4c}$$

where $f(E)$ again denotes the Fermi function. The above equation assumes that tunnel probability and electronic DOS can be taken as constant in the relevant energy range. This allows to evaluate the integrals with the help of

$$\zeta(x) = \int_{-\infty}^{\infty} dEf(E)(1-f(E-x)) = \frac{x}{\exp[x/k_B T]-1}. \tag{4.5}$$

4.4.3. Parameter Estimation

From First Excited State

In order to obtain a rough estimate of the parameters u, c_{ss} and c_{ts}, some simple calculations can be conducted. However, the results of this approach are expected to only give an idea of the order of magnitude.

When comparing the lifetime of the first excited state ($\tau_{\pm 7} \approx 500\,\text{fs}$) from eq. (4.1) to the mean time between two tunneling electrons at 1 nA which is of the order of 10^{-10} s, one can neglect the tunneling electrons for relaxation processes. However, the above lifetime value can just be taken as a lower limit. Therefore, when using, e.g., the transition[9] $|\Psi_7\rangle \to |\Psi_8\rangle$ to estimate the coupling to substrate electrons c_{ss}, its value must be considered an upper limit:

$$\tau_{|7|\to|8|}^{-1} \approx \Gamma_{7,8}^{S\to S} = \frac{c_{ss}}{4}\left|\langle\Psi_8\downarrow| J_+ |\Psi_7\uparrow\rangle\right|^2 \zeta(-E_{exc}) \tag{4.6a}$$

$$\Rightarrow c_{ss} \lesssim 7\times 10^{10}\,\text{meV}^{-1}\,\text{s}^{-1} \tag{4.6b}$$

[9]Transitions to $|\Psi_{-8}\rangle$ are possible but only with a very low probability, so that it does not affect this estimation.

If a future high-resolution experiment yields that $\tau_{\pm 7}$ is significantly higher, tunneling electrons may also have to be included in the relaxation rates.

For the estimation of c_{ts} and the spin-agnostic scattering u a slightly longer calculation is needed which relies on the excitation probability[10]. This is done by using the rates Γ to express the tunneling current and then comparing the elastic and inelastic contributions. It should be clear that c_{ts} depends on the tunneling conditions, i.e., the conductance set-point. Although the formulas for the determination are derived from the ITS situation, the interesting parameters are the ones in the spin-polarized measurements since for those conditions the lifetimes shall be calculated. In general, the current can be written as

$$I = I_{\text{el}} + I_{\text{inel}} = e\left[\sum_i N_i \left(\Gamma_{ii}^{T\to S} - \Gamma_{ii}^{S\to T} + \sum_{f\neq i}\left(\Gamma_{if}^{T\to S} - \Gamma_{if}^{S\to T}\right)\right)\right], \tag{4.7}$$

For V below the excitation threshold but sufficiently above zero, one tunneling direction, e.g., $S \to T$, can be neglected and then the current consists of the rates that leave the atom in its ± 8 state:

$$I(V) = e\left(\Gamma_{88}^{T\to S} N_8 + \Gamma_{-8-8}^{T\to S} N_{-8}\right) \tag{4.8a}$$

$$= e^2 V c_{\text{ts}} \left(\frac{1}{4}\left|\langle\Psi_8|\, J_z \,|\Psi_8\rangle\right|^2 + u^2 + u\eta \langle\Psi_8|\, J_z\,|\Psi_8\rangle (N_8 - N_{-8})\right) \tag{4.8b}$$

During the ITS experiments, unpolarized tips were used so that $\eta = 0$ which cancels the last term; also, $N_8 + N_{-8} = 1$ was used. This leads to the expression:

$$c_{\text{ts}} = \frac{I}{e^2 V (m^2 + u^2)}, \quad \text{where } m = \langle\Psi_8|\, J_z\,|\Psi_8\rangle/2. \tag{4.9}$$

In order to estimate c_{ts}, first the parameter u has to be determined. When turning to the situation of bias voltage above E_{exc}/e the current then consists of an additional inelastic contribution (see section 2.1.2, and fig. 2.4). Assigning an elastic and inelastic conductance leads to

$$I(V) = \begin{cases} c_{\text{el}} V & , eV < E_{\text{exc}} \\ c_{\text{el}} V + c_{\text{inel}}(V - E_{\text{exc}}/e) & , eV > E_{\text{exc}} \end{cases}. \tag{4.10}$$

The excitation probability (1 % to 5 %) can finally be identified as the ratio $c_{\text{inel}}/c_{\text{el}}$. Evaluating eq. (4.7) for the elastic and inelastic current then yields

$$c_{\text{el}} = e^2 c_{\text{ts}} \left(\frac{1}{4}\left|\langle\Psi_8|\, J_z\,|\Psi_8\rangle\right|^2 + u^2\right), \qquad c_{\text{inel}} = \frac{e^2 c_{\text{ts}}}{8}\left|\langle\Psi_7|\, J_-\,|\Psi_8\rangle\right|^2 \tag{4.11a}$$

$$\Rightarrow |u| \approx 5\ldots 14 \tag{4.11b}$$

$$c_{\text{ts}} \approx 6\times 10^6 \ldots 3\times 10^7 \text{meV}^{-1}\text{s}^{-1}, \quad \text{for } 5\,\text{mV}, 1\,\text{nA}. \tag{4.11c}$$

[10]It is assumed throughout this derivation that $B = 0$ and thus no direct transitions between the ground states are possible.

The spread comes from the variations in the excitation probability. This quantity is known to depend on the tip conditions, which is why only a range for both parameters can be given.

From Switching Probability

A second method to estimate the parameters c_{ss} and c_{ts} was attempted by a fitting procedure to the switching probability data from the voltage-pulse experiments. As explained in section 4.1 and shown in fig. 4.11, these experiments yielded a switching probability dependent on the pulse voltage. For almost all voltages, switching was attempted with roughly 100 pulses but the success rate stayed lower than 20 %, which is why the error bars are comparably large. The errors could not simply be computed by taking $\sqrt{N}$ of the number of successful switchings N, but were determined by using a binomial distribution instead, so that a) the lower bound does not go below zero, and b) the error bars span a 68 % confidence interval.

In the data, the onset of nonvanishing probability sits right at the excitation energy, above which the probability first rises to ~10 % and then almost saturates at around ~15 %. The shown fit curve comes from a stage of the master equation calculations when artificially introduced $|\Psi_{+6}\rangle$ and $|\Psi_{-6}\rangle$ were used instead of the $\mathcal{H}_{CF}$ eigenstates, but only this way the experimental data could be described reasonably well. The fit curve also turned out to depend sensitively on η but only very weakly on the coupling values $c_{ss/ts}$. Rather, different values spanning several orders of magnitude produced almost the same curve, as long as their ratio c_{ss}/c_{ts} was the same. Thus, it must be accepted that this data set does not allow to accurately determine the absolute value of the coupling parameters. Nevertheless, the estimated values of $c_{ss} \approx 4 \times 10^4\,\text{meV}^{-1}\,\text{s}^{-1}$ and $c_{ts} \approx 6 \times 10^4\,\text{meV}^{-1}\,\text{s}^{-1}$ can be seen as lower limits, below which an acceptable agreement between calculations and experiment could not be achieved any more.

The comparison of the two methods to determine the coupling parameters showed that their outcomes differ considerably by several orders of magnitude. Furthermore, when using the eigenstates of $\mathcal{H}_{CF}$, the calculated switching probability always showed some discrepancy to the data, because upon reaching the excitation energy of the $\left|\Psi_6^{a/s}\right\rangle$ a strong increase of the switching probability was predicted from the calculations that was not observed in the experiment. Rather this false prediction could be explained by the 50 % probability of the system to relax to one of the ground states when the $\left|\Psi_6^{a/s}\right\rangle$ were populated. However, it was found that due to (previously ignored) decoherence, induced by the tunneling electrons, these states undergo yet another reshaping and return to states close to the parabola with $|\langle J_z\rangle| \approx 6$. In the quantum master equation approach this was found to be a voltage-dependent effect [36] with substantial influence on the shape of the whole spectrum.

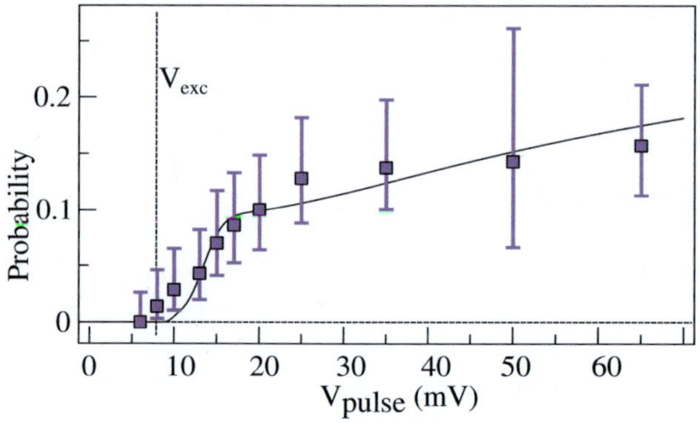

Figure 4.11.: *Switching probability with voltage pulses. The duration of each pulse was* 200 µs *and its amplitude was varied up to* 65 mV. *Error bars result from a finite number of attempts and indicate one standard deviation. Parameters for the fit (black line) are* $\eta = 0.19 \pm 0.02$, $c_{ss} = 4 \times 10^4\,\mathrm{meV^{-1}\,s^{-1}}$, *and* $c_{ts} = 6 \times 10^4\,\mathrm{meV^{-1}\,s^{-1}}$.

Not only did the two parameter estimates yield different results, but also the predictions of the rate equation calculations never truly succeeded to consistently describe the experimental data (switching behavior and ground state lifetimes) with one set of parameters. Taking temperature and modulation voltage broadening into account did not improve the situation substantially, or could only approach an approximate agreement for unrealistically high temperatures which could be ruled out from experiment. The final conclusion from the failure of the classical rate equation approach, based on the spectrum of $\mathcal{H}_{CF}$, was that a full quantum master equation calculation is needed in order to capture the dynamics of this system.

4.4.4. Tunneling Current and Temperature

It is not surprising that the measurements of the dynamics of the Ho/Pt(111) system show indications of two additional dependencies, namely tunneling current and temperature. Since electrons and their role in the scattering process are at the heart of the explanation, their number and energy distribution are obvious quantities that need to be considered when predictions about the ground state lifetimes are made. Partially this was mentioned already in section 4.2.2 and 4.2.3. Although the rate equation approach did not succeed to quantitatively describe the experimental data, at least some qualitative expectations for current and temperature could be formulated and tested.

Calculating the transition rates between the ground states from eq. (4.2)-(4.4) with the assumption of a low magnetic field so that transitions are weakly allowed ($\Delta E_Z \ll k_B T$) and also low temperatures ($k_B T \ll E_{exc}$), yields approximately that $\Gamma_{-8,8} \approx \Gamma_{8,-8} \propto k_B T$.

In other words, this rough estimate predicts that the lifetimes should be inversely proportional to temperature or, equivalently, that the product of mean lifetime and temperature is roughly constant. The experimental data allows only to compare two series of measurement: the already presented lifetime at 5 mV, 50 nA, $B = 0$, and 700 mK and another result from the same sample under identical conditions except for temperature (4.4 K). In the latter case, the analysis of the upper and lower states gave a combined mean lifetime of 4 s, the former yielded ~20 s. The two data sets produce an acceptable agreement:

$$4.4\,\mathrm{K} \cdot \tau_{4.4\mathrm{K}} = 17.4\,\mathrm{K\,s}, \qquad 0.7\,\mathrm{K} \cdot \tau_{0.7\mathrm{K}} = 14\,\mathrm{K\,s} \tag{4.12}$$

If the distinct upper and lower state values are considered, the lower states agree well at $T \cdot \tau_{low} \approx 8\,\mathrm{K\,s}$ and the upper ones have a similar value at $T \cdot \tau_{up} \approx 22\,\mathrm{K\,s}$. Of course, many factors were neglected in this estimate, e.g., the spin-polarization and energy of the tunneling electrons, but the fair agreement indicates that qualitatively the rate equation approach may still give reasonably accurate predictions.

A quantitative estimate of the influence of the tunneling current is more difficult and faces the problems of little-known coupling strengths to tunneling and substrate electrons mentioned above. Especially, it is unclear at which current the tunneling electrons dominate the switching rates. Therefore, a clear statement about the scaling of lifetimes with current cannot be given, except for the expectation that higher currents should lead to shorter lifetimes. Moreover, the effect of a changed current cannot be separated from the influence of the tip which is at a different distance.

4.5. Quantum Master Equation

In [36], Karlewski et al. theoretically studied the dynamics of *fcc* holmium adatoms on Pt(111) using a full quantum master equation approach, and the *ab initio* CF parameters from [9]. The main motivation behind it was the realization that the dissipation caused by the tunneling electrons induces decoherence and leads to significant changes in the spectral composition of some of the states. The eigenstates of $\mathcal{H}_{\mathrm{CF}}$ can therefore not be used without restrictions. Instead, the correct basis has to be taken from the steady-state solution of the density matrix for a given bias voltage. This is particularly important for the $\left|\Psi^{a/s}_{6/3}\right\rangle$ states, since their phase relation is destroyed, but also under symmetry-breaking conditions such as a B_2^1 term or a magnetic field, the eigenstates of $\mathcal{H}_{\mathrm{CF}}$ are much more prone to forming nonmagnetic linear combinations.

Two examples of the steady-state basis are compared with the spectrum of $\mathcal{H}_{\mathrm{CF}}$ in fig. 4.12 for different bias voltages. A minute magnetic field of $B_z = 1 \times 10^{-8}\,\mathrm{T}$ was included in the calculations, and the couplings[11] amount to $c_{\mathrm{ts}} = 3.4 \times 10^6\,\mathrm{meV^{-1}\,s^{-1}}$, $c_{\mathrm{ss}} = 0$. The influence of c_{ss} on the spectrum was found to be weak, even when it is greater than c_{ts}.

[11] The same notation as in section 4.4.2 is used.

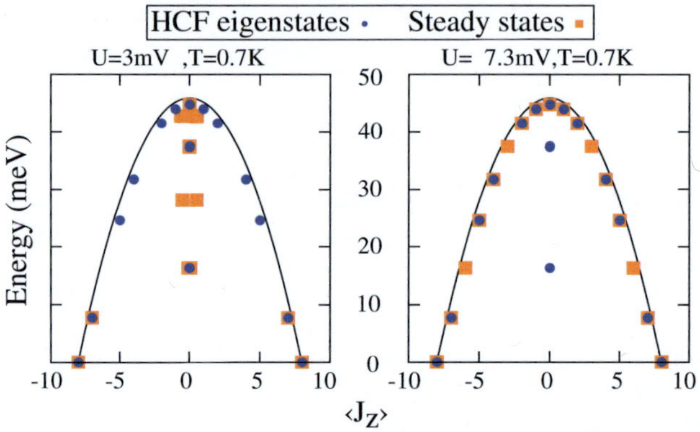

Figure 4.12.: *Comparison of $\mathcal{H}_{CF}$ eigenstates with the bias-dependent steady-state basis. While at* 3 mV *the steady-state calculations presumably suffer from numerical inaccuracies in the upper part of the spectrum, predicting mostly nonmagnetic states, the dissipation at higher bias voltage (*7.3 mV*) restores the twofold degeneracy for all states. However, higher excited states are negligible at low voltages since they cannot be populated, but will show the correct behavior once the bias is high enough. In contrast, the $\mathcal{H}_{CF}$ eigenstates do not change at all with bias and still exhibit the* $\left|\Psi_6^{a/s}\right\rangle$ *that lead to increased "shortcut" switching between the ground states at sufficient voltages.*

In the idealized system with no symmetry-breaking except for the minute B-field mentioned before, the calculations showed that the lifetimes decrease exponentially with higher bias voltage for $V \lesssim E_{78}$ (see fig. 4.13) due to finite temperature ($T = 0.7\,\mathrm{K}$). Above that value, the lifetimes first decrease even faster but then saturate at roughly 9 mV since all possible excitations can be reached, and thus all possible switching paths will be taken. As was suggested already in the comment on the lifetime of the first excited state, its value is not well known and could actually be significantly higher than the estimate in eq. (4.1). This is true in the calculations as long as $c_{ss} = 0$. Therefore, the tunneling electrons might find the adatom in its first excited state and even populate higher states, which finally enables all excitations across the barrier once the system's largest splitting $\Delta E_{67} = E_{\pm 6} - E_{\pm 7}$ can be overcome. However, the low-voltage lifetimes are many orders of magnitude too high ($\tau \approx 10^{20}$ s) since some important correction terms are still missing. Furthermore, the temperature dependence is extremely strong in the idealized calculation, which is in contrast to the findings in the previous section.

When considering the modulation voltage $V_{mod} = 0.8\,\mathrm{mV}$ one finds that effectively the bias voltage is increased so that the lifetime curve is merely shifted towards lower voltages. Also, adding a nonzero coupling to the bath electrons c_{ss} hardly changes the computed lifetimes; if it is significantly larger than c_{ts} the lifetimes are actually enhanced (not shown). This is because now the relaxation from the first excited state back into the ground state induced by substrate electrons will inhibit further excitations to higher states.

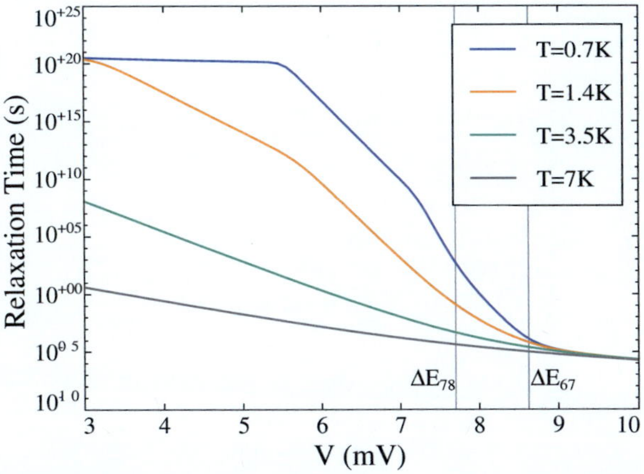

Figure 4.13.: *Calculated dependence of ground state lifetimes on voltage. The lifetimes span many orders of magnitude in this idealized scenario, the coupling parameters are* $c_{ts} = 3.4 \times 10^6\,\mathrm{meV^{-1}\,s^{-1}}$, $c_{ss} = 0$, *and* $B_z = 10^{-8}\,\mathrm{T}$. *The low-energy plateau for T=0.7 K is due to this minute field. Four different curves illustrate the rapid decrease of lifetimes with increasing temperature.*

Only by including a symmetry-breaking B_2^1 and adding an artificial electronic broadening γ of almost 1 mV, the ground state lifetimes from the bias-dependence experiment (section 4.2.3) can be approached within one order of magnitude (see fig. 4.14). Furthermore, an asymmetry of the two ground state lifetimes is found due to the spin-transfer torque effect (with moderate spin-polarization $\eta = 0.15$). Still, it is not certain that only this parameter set is suitable to achieve proper agreement with the experimental data.

Of course, also the lifetime dependence on magnetic field (in z-direction) was investigated in the calculations. For the idealized system, already weak magnetic fields drastically reduce the lifetimes by about ten orders of magnitude for less than 50 mT (from 10^{20} s to 10^{10} s), and then a much slower decrease towards higher fields is found. Hence, the experimental results can be qualitatively described with the quantum master equation if a weak background field is assumed. However, when including the broadening γ and a B_2^1 of 4×10^{-4} meV the already very short lifetimes (merely seconds) are enhanced with higher fields. The interpretation in this case would be that the magnetic field partially counteracts the mixing by O_2^1 and helps to preserve states with finite $\langle J_z \rangle$. Whether this is a likely scenario that is realized in the experimental situation is yet unclear.

Like the rate equation approach before, also these quantum master equation calculations suffer from the scarcity of experimental data for better determination or testing of the various parameters. However, this approach is anticipated to give more accurate predictions than a classical rate equation calculation and may help to identify the interesting ranges for experimentally variable parameters in order to further test the interpretation.

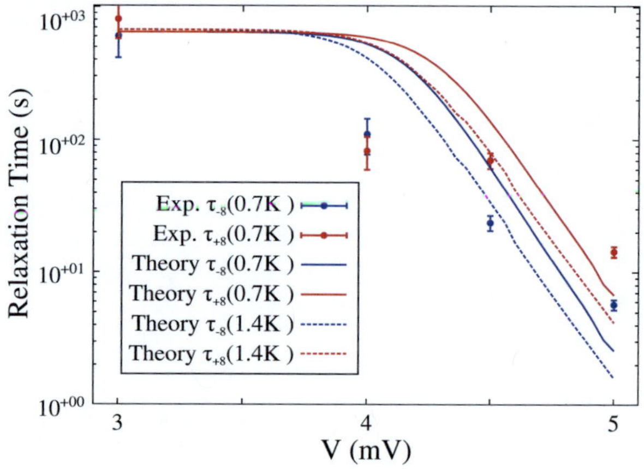

Figure 4.14.: *Comparison of experimental with calculated ground state lifetimes when the tip's spin-polarization η, a modulation voltage V_{mod}, a symmetry-breaking operator O_2^1 and an additional electronic broadening γ (see text) are included (but still with $B_z = 10^{-8}$ T). Curves for two different temperatures (0.7 K and 1.4 K) are shown. Similar lifetimes as from the experiment and even a lifetime asymmetry can be obtained, only the voltage dependence does not yet agree perfectly.*

4.6. Miscellaneous

As a last topic, further magnetic interactions of a Ho adatom will briefly be discussed. These interactions cover the hyperfine coupling to the atom's nucleus, as well as indirect interactions with surrounding magnetic structures, such as Co islands and other Ho adatoms via the RKKY-mechanism. Furthermore, the symmetry considerations from the simple CF Hamiltonian model are revisited and systematically applied to all other $4f$ elements. This is done in order to identify other combinations of adatom and adsorption site symmetry that would allow stable ground states similar to Ho on Pt(111).

4.6.1. RKKY Interaction

The RKKY interaction [46–48] is an indirect coupling of two magnetic moments, or of one moment to an extended magnetic structure via the polarization of substrate electrons at the Fermi edge. It is therefore a higher order process and involves at least two exchange-like scattering events with the electron bath. Two studies of cobalt atoms on a Pt(111) surface investigated the RKKY interaction between Co islands at the step edges to Co atoms [42], and the coupling between pairs of atoms at different relative positions [49].

In the isotropic case, the interaction of two angular momenta $\vec{S}_1$ and $\vec{S}_2$ at a distance d can be described as an exchange-like Hamiltonian with an oscillating and distance-dependent coupling strength that leads to alternating ferromagnetic and antiferromagnetic coupling:

$$\mathcal{H}_{\text{RKKY}} = J(d)\vec{S}_1 \cdot \vec{S}_2, \quad \text{where } J(d) \propto \frac{\cos(2k_F d)}{(2k_F d)^D}. \tag{4.13}$$

The exponent D depends on the dimensionality of the system. For the case of two adatoms this would give $D = 2$; for the coupling of one atom to a Co island or stripe the expected dimensionality is close to one, so that $D \approx 1$. Of course, in the latter case also one of the $\vec{S}$ has to be replaced by an effective exchange field. In [49], the authors found that the above Hamiltonian is only valid along the same direction, i.e., their results indicate that the coupling J is also direction-dependent.

The consequences of RKKY interaction on the holmium case are rather limited, though. As long as no external magnetic field is present, the protection from time-inversion symmetry still holds and forbids transitions between the ground states. For that reason, other magnetic adatoms do not directly lead to higher transition rates. The matrix element $\langle\Psi_M| J_z |\Psi_M\rangle$, however, is nonzero for most M and thus the RKKY interaction can lead to an energy shift of the levels.

4.6.2. Hyperfine Interaction

Holmium is known to have only one stable isotope (^{165}Ho) with a nuclear spin of $I = 7/2$, which leads to eight different nuclear levels. Usually, the hyperfine interaction couples nuclear and electronic total angular momentum, giving $\vec{F} = \vec{J} + \vec{I}$ which would result in a half-integer $F = 23/2$. If a quadrupole term is present the simplified[12] Hamilton operator can be written as

$$\mathcal{H}_{\text{hfs}} = A\left(\vec{J} \cdot \vec{I}\right) + \frac{3B}{4I(2I-1)J(2J-1)}\left[\left(\vec{J} \cdot \vec{I}\right)^2 + \frac{3}{2}\vec{J} \cdot \vec{I}\right], \tag{4.14}$$

where $A = 3.3 \times 10^{-3}$ meV and $B = -6.89 \times 10^{-3}$ meV [50] are the coupling strengths for the dipolar and the quadrupolar contributions — though for the free, neutral atom. As in the case of an exchange interaction, the dipole term can only cause a shift of the levels but not lead to increased ground state transitions. The quadrupole term which contains operators with $J_\pm^2 I_\mp^2$, however, can induce extra transitions. The matrix elements are found to depend sensitively on the external magnetic field, though, which partially counteracts the effect of the quadrupole term. The lowest two octets which are approximately the $|\Psi_{\pm 8}\rangle \otimes |I_z\rangle$-states are shown in fig. 4.15.

Since the splitting due to $\mathcal{H}_{\text{hfs}}$ is comparable to $k_B T$, all states from the two lowest octets are expected to be at least partially populated at 0.7 K. Therefore, all 16 states have to

[12] without constant terms

be considered when studying the dynamics. Yet, the calculation of the matrix elements only yields noteworthy values for transitions among the states with the same $|\Psi_M\rangle$ with $\Delta\langle F_z\rangle \approx 1$. All other matrix elements are very small or nonzero only due to numerical inaccuracies. Likewise, when the influence of magnetic field is calculated, the different levels do not show any avoided crossing larger than ~ 0.1 neV, which means that there is no significant coupling between the ground state octets on both sides of the barrier.

Lastly, it is yet unclear whether the influence of the electron bath is strong enough to decouple nuclear and electronic angular momentum. Quantum master equation calculations with hyperfine interaction are planned and likely to be carried out in the near future.

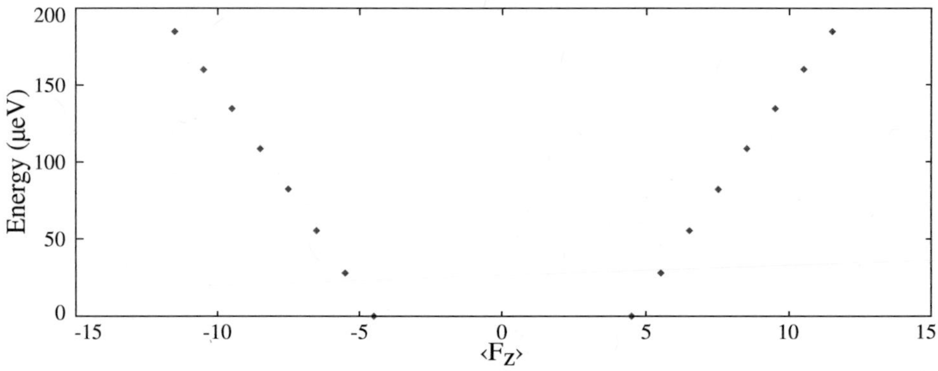

Figure 4.15.: *Ground state octets forming due to the hyperfine interaction with A, B as given in the text and $B_z = 10^{-8}$ T. The splitting is less than* 200 μeV.

4.6.3. Prediction for Rare Earth Elements

Despite the difficulties regarding the calculations, the fundamental arguments of symmetry-protected ground states were never weakened and still hold. For this reason, a step towards a broader prediction based on the initially used concept of crystal field Hamiltonians may be dared by analyzing all possibilities of rare-earth adatoms in various adsorption site symmetries. From the discussion of magnetic adatoms so far and the desire to store information in their magnetization, it should come as no surprise that in the following only scenarios with out-of plane easy axis anisotropy will be assumed since two degenerate ground states with high $\langle J_z\rangle$ are sought-after. Still, finding proper combinations of substrate and element for which this assumption holds remains a nontrivial task. The following discussion can, however, serve as a starting point for such considerations.

Table 4.1 presents the expected aptitude — or lack thereof — of stable magnetism in the ground states. The available symmetries are limited to those relevant for adsorption on surfaces[13]. It is found that promising systems are systematically outnumbered: Both twofold and fourfold symmetric sites cannot allow stable magnetic ground states for the assumed $4f$ configurations of rare-earth atoms. In these adsorption sites, adatoms will either share the fate of the $J = 2$ example from section 3.1.3, having $\mathcal{H}_{CF}$ couple the $|\pm J\rangle$ to form a single nonmagnetic ground state, or the system is susceptible to Kondo-singlet formation. Above the Kondo temperature T_K scattering will still lead to magnetization reversal since the ground state transitions are not forbidden.

Fortunately, there remain several promising combinations: The interesting systems can thus be found in three- or sixfold symmetries. Systems with integer J can be stable up to second-order processes, with the restriction that J must not be an integer multiple of 3. In some half-integer systems even greater stability is predicted. These constellations have the additional protection of Kramers theorem so that their spectrum must be at least twofold degenerate, even when the spatial symmetry does not immediately suggest this. Therefore, when $2J = 3n, \; n \in \mathbb{N}$, consecutive scattering with at least three electrons is needed to reverse the magnetization. All other half-integer J should be Kondo cases, but there are two exceptions — cerium and samarium in sixfold symmetric sites — for a trivial reason: The constraints on the allowed Stevens operators limit $\mathcal{H}_{CF}$ so severely that no sixth-order terms appear ($2J < 6$) and thus all eigenstates remain pure J_z-eigenstates. In other words, no mixing occurs and there can be no "tunneling of magnetization" through the anisotropy barrier, it must be overcome completely by five successive scattering events.

In conclusion, holmium on Pt(111) seems to be the most stable integer-J rare-earth system, since the anisotropy barrier and therefore E_{exc} should be expected higher than for the $J = 4$ cases. For even greater stability, half-integer systems, and especially the 5/2-systems in C_6 should be investigated. Of course, all this presumes that such a system can be implemented and that its properties are experimentally accessible.

	La^{3+}	Ce^{3+}	Pr^{3+}	Nd^{3+}	Pm^{3+}	Sm^{3+}	Eu^{3+}	Gd^{3+}	Tb^{3+}	Dy^{3+}	Ho^{3+}	Er^{3+}	Tm^{3+}	Yb^{3+}
J	0	5/2	4	9/2	4	5/2	0	7/2	6	15/2	8	15/2	6	7/2
C_2	-	K	S	K	S	K	-	K	S	K	S	K	S	K
C_3	-	K	2	3	2	K	-	K	S	3	2	3	S	K
C_4	-	K	S	K	S	K	-	K	S	K	S	K	S	K
C_6	-	5	2	3	2	5	-	K	S	3	2	3	S	K

Table 4.1.: *Prediction of stability of the ground states of adatoms (3+) from the lanthanide elements in an adsorption site with given symmetry. It is assumed that they exhibit an out-of plane easy axis, such that the ground states have maximum $\langle J_z \rangle$. As before with Ho (and Gd), J is based on Hund's rules. The letter K marks the Kondo-case, an S indicates a non-magnetic singlet ground state, and the numbers 2, 3, 5 signify the lowest-order electron scattering process that can lead to magnetization reversal. Note that for $J = 0$ no prediction can be made, since from the $4f$-shells L and S are nonexistent or expected to cancel.*

[13] It should not need any mention that a C_1 symmetry is utterly unsuitable.

4.7. Open Questions

Alternative Interpretations of Experimental Data

Scanning tunneling spectroscopy techniques can usually only determine excitation energies and probabilities. The identification of the nature of the inelastic signatures in the curves must rely on additional information from experiments with varied parameters or from theoretical considerations. According to the *ab initio* calculations for Ho/Pt(111), the adatoms should exhibit a vibration mode with an energy of $\sim$5 meV. This excitation could in principle lead to a very similar dip-peak signature in d^2I/dV^2 measurements. Its energy should remain unaffected by an external magnetic field, as is also briefly discussed in the next paragraphs, which can be used to confirm or exclude this type of excitation. Furthermore, spin excitations could in principle also show different inelastic intensities in positive and negative bias polarity if a spin-polarized tip is used [34, 44]. However, since both ITS and sp-STM are experimentally challenging techniques, they are rarely used in a combined fashion and the expected success rate is not very high. Already scheduled ITS experiments with magnetic field should therefore be sufficient to distinguish between vibrational and spin excitations.

An STM study on adatoms from the lanthanides (La and Ce) by Pivetta et al. [51] observed different topographic appearance upon the co-adsorption of hydrogen. Also the spectroscopic signature changed around zero bias, showing a broad excitation-like step in the differential conductance. This was attributed to vibrations of the adatom with one hydrogen atom sitting "below" the adsorption site. Hence the question arises if the observed ITS signal and switching behavior of holmium could originate from such "hydrogen-poisoning". In contrast to the experiments by Pivetta et al. no such topographic contrast was ever observed in the Ho/Pt(111) experiments. Moreover, the magnetic fields used in the lifetime dependence measurements (section 4.2.2) are by far too weak to have such a marked influence on the switching time of conformational changes. Thus, all kinds of vibrational excitations can be considered highly improbable.

Conflicting XAS/XMCD Experiments

In a recent experiment by Donati et al. [52] Ho adatoms on Pt(111) were studied with X-ray absorption spectroscopy (XAS) and X-ray circular magnetic dichroism (XMCD). Their experiments could confirm the ground state multiplet as $J = 8$, however, further multiplet calculations did not lead to a similar anisotropy barrier as in the case presented in this thesis. Rather, the best fit to their experimental results was obtained with an anisotropy that has its minima at the $|\Psi_{\pm 6}\rangle$ states. In the absence of decoherence this would again lead to a symmetric and antisymmetric linear combination and result in a single nonmagnetic ground state. Furthermore, magnetization curves did not reveal any hysteretic behavior which is why the authors express their doubts about the long lifetimes of the ground states.

This discrepancy between the experiment by Donati et al. and the combined experimental and theoretical study presented in this thesis cannot be resolved, yet. A measurement of the Zeeman effect on the transition energy to the first excited state should reveal $|\langle J_z \rangle|$ of the ground states. From theory, a Zeeman shift of $72\,\mu\text{eV}\,\text{T}^{-1}$ is expected, so that the measurements need good energy resolution and high fields. Although the JT-STM has a 3 T coil, the current implementation has some weaknesses, so that its maximum field has been limited to only $\sim$1.6 T. The new d-STM reaches lower temperatures as well as higher fields and thus should allow to measure this effect — the measurements are currently under preparation and are going to be carried out in the near future. If the calculations of Donati et al. are correct then actually two excitations should be observed: from $|\Psi\rangle_{\pm 6}$ to $|\Psi\rangle_{\pm 5}$ or $|\Psi\rangle_{\pm 7}$. Furthermore, the effect of magnetic field would shift them in opposite directions.

A second issue concerns the averaging character of XAS/XMCD — most probably, Donati et al. measured the average response of adatoms in *fcc* and *hcp* positions in an extended area as both adsorption sites should be populated. Since the *ab initio* calculations predict different anisotropy constants for the *hcp* site, their multiplet calculations cannot necessarily unite both adsorption sites and still accurately describe the *fcc* case. Again, highly resolved ITS measurements of both species could help to explain the discrepancy, since improved ITS measurements on *hcp* adatoms will allow to assess the *ab initio* CF parameters. For the time being, the good agreement of the *fcc* case leads to the expectation that the *hcp* calculations are accurate, as well.

Chapter 5.

5 Exploratory Measurements on Other Single Atoms

Besides the holmium investigations, two other experiments to measure spin excitations in single adatoms were conducted. The first project was inspired by the Ho findings, since again a 4f element, cerium (Ce), was the material of choice. Although the surface was an insulating copper nitride on Cu(100) layer that does not offer one of the promising symmetries for stable magnetic moments, this project can be seen as another test of the accessibility of the magnetic properties of 4f adatoms with STM. In another experiment, cobalt (Co) was investigated on an insulating film of copper nitride on the anisotropic Cu(110) surface. Here, the goal was to find magnetic excitations in single atoms and subsequently investigate the excitation spectrum of self-assembled chains. Since both systems involve a different substrate than before, a brief description of the surfaces will be given in the respective sections.

5.1. Cerium on Copper Nitride

Cu_2N on Cu(100) and Adsorption Sites

Already the first observation of spin excitations with STM [6] was achieved on an insulating Al_2O_3 layer on a metallic substrate. Subsequently, more such measurements were published (e.g., [38, 39, 44, 53, 54]), albeit with a thin, one-atomic layer of copper nitride (Cu_2N) on Cu(100) which has become one of the standard materials for the search of spin excitations in atomic-scale adsorbates.

This surface was investigated prior to the above measurements for the first time by Leiblse et al. [55, 56] in the 1990s. Depending on the sample preparation, either – at

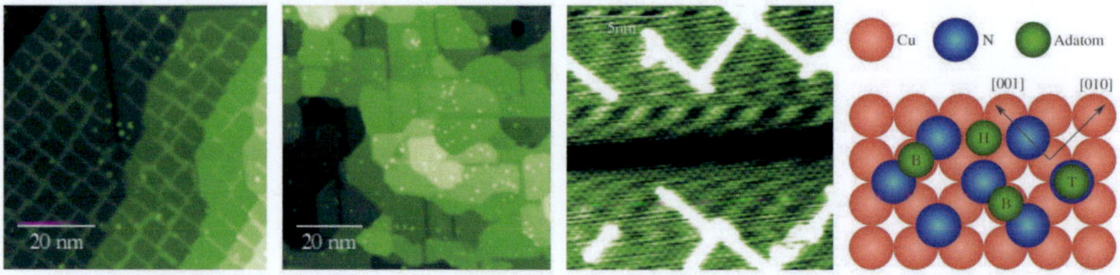

(a) (b) (c) (d)

Figure 5.1.: *(a) Topography of partial and (b) full coverage of the Cu(100) surface with copper nitride. Both surfaces are already decorated by adatoms and clusters, visible as bright round protrusions. (c) Atomically resolved image of the Cu_2N surface and (d) schematic drawing of the adsorption sites of adatoms on Cu_2N/Cu(100): The nitrogen sites are marked in blue, and the three high-symmetry sites are indicated in green. The letters "B", "T", and "H" denote the bridge position, top, and fourfold hollow site.*

low coverages – copper nitride islands of roughly 5×5 nm^2 form, arranged in an approximately rectangular grid, or the copper surface is completely covered with a Cu_2N overlayer which reduces its slight lattice mismatch by forming long trenches along the $\{011\}$ directions. Both structures are depicted in fig. 5.1. The islands in fig. 5.1(a) appear as depressions in the topography since the LDOS is higher in copper.

Referring to the surface unit cell of bare copper, Cu_2N grows in a c(2×2)-structure, with nitrogen occupying every other fourfold hollow site between the copper atoms. The nitrogen positions can be determined by atomically resolved imaging which shows the atoms as small protrusions (see fig. 5.1(c)). Single atoms from the transition metals (e.g., [38, 39]) were found to adsorb on this structure mostly in the "bridge" positions (fig. 5.1(d)). So far, no studies reported on the adsorption site of rare earth atoms on Cu_2N. Yet, data of Schuh [10] suggest that holmium can be found in the "top" position. Whether this can be generalized to all *4f* elements remains to be seen. For the case of cerium, the adsorption site(s) had to be identified since it determines the structure of $\mathcal{H}_{\mathrm{CF}}$: The bridge position exhibits a $C_{2\mathrm{v}}$ symmetry, whereas the two other possible adsorption sites – the top position and the fourfold-hollow site – both possess a $C_{4\mathrm{v}}$ symmetry.

Spectroscopy of Cerium on surfaces

Cerium single atoms were first studied with STM on a Ag(111) surface [37] where a Kondo resonance at the Fermi level was found. However, subsequent studies with Ce on Ag(100), Cu(100) and even Cu_2N/Cu(100) did not reveal any Kondo-like features in the conductance spectra[1] [57, 58]. Rather the appearance of pseudo-Kondo resonances due to the presence of hydrogen was reported [58], casting doubts on the original findings.

[1] Although experiments for Ce on Cu_2N/Cu(100) are mentioned in [57], a publication of neither qualitative nor quantitative results could be found.

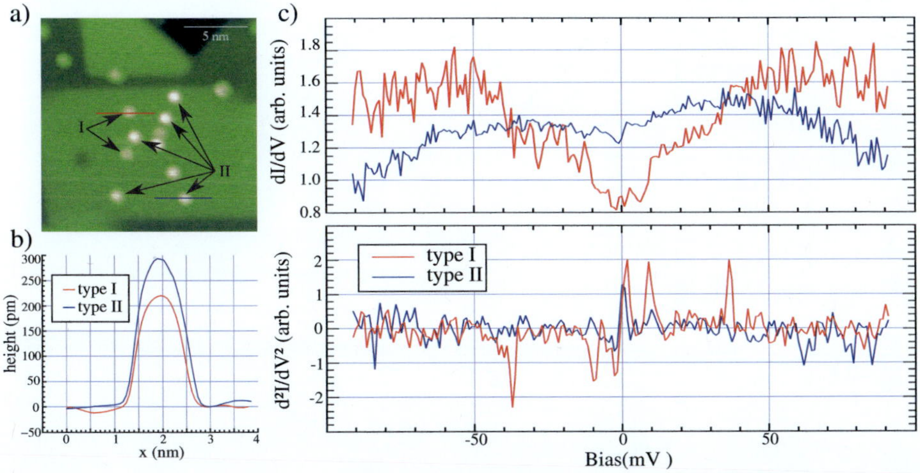

Figure 5.2.: *(a) The topography reveals two different species of Ce atoms with an apparent height of* $\sim 220\,\mathrm{pm}$ *(type I) and* $\sim 300\,\mathrm{pm}$ *(type II);* $V = 200\,\mathrm{mV}$, $I = 50\,\mathrm{pA}$. *The red and blue lines correspond to the curves in the cross-section plot (b). (c) The spectra of type I and II differ considerably: type I shows three excitations at* $\sim 2, 10$, *and* $36\,\mathrm{meV}$, *while type II hardly shows any excitation. Experimental parameters:* T=880 mK, I=2 nA *at* 90 mV, $V_{\mathrm{mod}} = 0.8\,\mathrm{mV}$, $\vec{B} = 0$.

Moreover, Ce atoms were found to be very mobile on the metallic surfaces even at low temperatures of $\lesssim 10\,\mathrm{K}$ [59]. Hence it is uncertain whether actual Ce single atoms were investigated in the first experiments. Due to these uncertainties and the desire to study spin excitations in cerium single atoms, a new series of measurements of Ce atoms on the Cu_2N/Cu(100) surface was started during this thesis.

After deposition of Ce to the surface mainly two species of atomic-sized adsorbates (which will be called type I/II below) were found with different heights, as is shown in fig. 5.2(a) and (b). Furthermore, the two types were also found to exhibit different tunneling spectra (see fig. 5.2(c)). Although these are not high-quality spectra they clearly illustrate the presence of excitations in only one adsorbate type. No qualitative difference could be found between type II spectra and those taken on Cu_2N (not shown). Type I adsorbates show three distinct excitations at $\sim 2, 10$, and $36\,\mathrm{meV}$, clearly identified as dip-peak pairs in the d^2I/dV^2 curves and as conductance steps, slightly obscured by noise, in the dI/dV spectra. The results of higher quality spectroscopy measurements are presented and discussed below.

After finding these excitations, subsequent measurements mostly focused on type I adsorbates, in order to obtain higher-resolution spectra, investigate a potential magnetic field dependence of the excitations, and reveal their adsorption site. Prior to that, some spectra with a wider bias range ($\pm 300\,\mathrm{mV}$) were taken in order not to miss any possible higher excitations. The spectra shown in fig. 5.3 were subject to (anti-)symmetrization

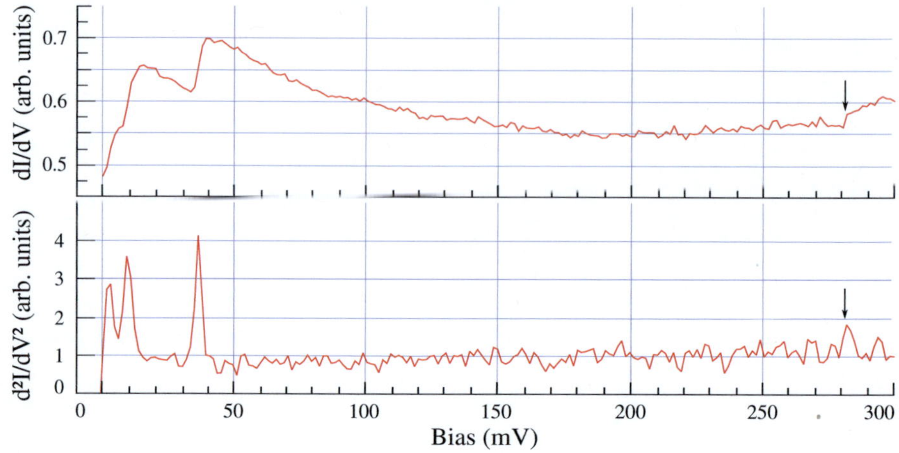

Figure 5.3.: *The spectra covering a larger bias range show only one very weak hint of an excitation at* $\sim$ 282 meV *(arrow) in addition to the already known low-energy excitations. The plot shows spectra after (anti-)symmetrization. Experimental parameters:* T=840 mK, I=5 nA *at* 300 mV, $V_{\mathrm{mod}} = 1.2\,\mathrm{mV}$, $\vec{B} = 0$.

in order to improve the signal quality of common features at positive and negative bias polarity. Besides the three low-energy excitations, another faint conductance step is visible at $\sim$ 282 mV, just below the upper voltage limit of the spectrum. A corresponding peak can be found in the d^2I/dV^2 curve, barely strong enough to be distinguished from the noise in the spectrum. As a preliminary interpretation, this excitation is assumed to be a transition to the next multiplet, as for several Ce oxidation states such excitations occur between 270 meV and 300 meV [50]. An exact conclusion cannot be drawn, yet, but further measurements should allow to identify the observed transition.

Spectra taken with better resolution than shown before were fitted to extract the excitation energies. This was done for measurements at B=0 and B=1 T. The results are summarized in fig. 5.4. A shift of a few hundred microvolts towards higher energies is found for all three peaks at B=1 T. While the 95 % confidence interval for ΔE_2 is larger than the value itself, the shifts ΔE_1 and ΔE_3 can be considered statistically significant. For magnetic excitations, however, the observed shifts are rather large. In order to identify the corresponding transitions, more data points at different fields would be needed to study the progression of $\Delta E(B_z)$. Unfortunately, only these two spectra could be taken, and higher fields were not possible during that particular experiment.

Since, at least until now, no results on magnetic properties of cerium single atoms on Cu_2N/Cu(100) have been reported, the oxidation state and consequently J are also unknown. Nevertheless, some of the above results may be used as arguments for reasonable speculation. For several transition metals on this substrate it is known that those atoms release their two s-electrons (e.g., [38]) but keep the d-shell as in the free atom. Adapting this behavior for cerium results in a +2 oxidation state. This state has a configuration of

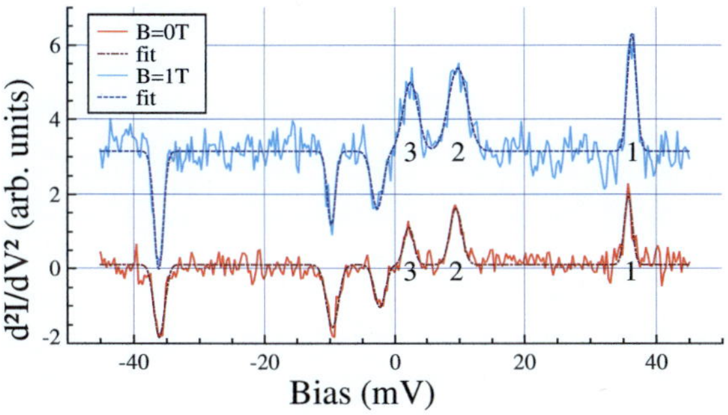

	B=0T	B=1T
E_1	35.85 ± 0.04	36.25 ± 0.04
E_2	9.55 ±0 .05	9.74 ± 0.07
E_3	2.18 ± 0.06	2.56 ± 0.09
ΔE_1	0.40 ± 0.08	
ΔE_2	0.19 ± 0.12	
ΔE_3	0.38 ± 0.15	

Figure 5.4.: *Inelastic spectra and fitted Gaussians to determine the excitation energies at B*=0 T *and B*=1 T. *All energies are given in* meV. *The spectra at are vertically offset for clarity. Experimental parameters:* $T \approx 0.9$ K, *I*=1.5 nA *at* 45 mV, $V_{\mathrm{mod}} = 0.6$ mV, 3.07 kHz.

[Xe]4f^2, according to [50], i.e., the $5d$-electron is transferred to the f-orbitals. Thus, using Hund's rules yields $L = 5, S = 1$, and $J = 4$ with $g_J = 0.8$. With these properties it is easy to explain three excitations in a $C_{2v/4v}$ symmetry due to the mixing operators in $\mathcal{H}_{CF}$. The resulting eigenstates would be non-magnetic since they would have $\langle J_z \rangle \approx 0$.
Two arguments oppose this scenario: First, if the observed excitation at ~280 meV is interpreted as a transition to the second lowest multiplet, then the $^3H_4 \rightarrow {}^3H_5$ transition of the Ce^{2+} ion does not fit to the measured energy. This transition is expected approximately 100 meV lower than the observed one. Second, according to [60] the +2 oxidation state is very rarely found and only in very specific solids[2]. It seems questionable whether Ce is present in this configuration on this particular surface.

In contrast, if one assumes the typical lanthanide behavior in compounds, the +3 oxidation state could be hypothesized. This state has a [Xe]4f^1 configuration, leading to a $^2F_{5/2}$ ground state multiplet and $g_J = 6/7$; the transition to $^2F_{7/2}$ lies at 279 meV [50] and agrees very well with the measured spectrum. However, for a $J = 5/2$ system in $C_{2v/4v}$ it is impossible to obtain three excitations within the same multiplet for B=0. If the adatoms are actually Ce^{3+}, then at least one of the low-energy excitations must be of non-magnetic origin. Since excitations at very low energies, i.e., directly at the Fermi level, appear rather frequently[3], one could tentatively exclude the measured excitation (3) from the analysis in order to estimate crystal field parameters. The ground states would then have to be $|\Psi_{\pm 3/2}\rangle$ to allow for two excitations of approximately equal intensity. Assuming a C_{2v} symmetry, the energies of $|\Psi_{\pm 1/2}\rangle$ and $|\Psi_{\pm 5/2}\rangle$ then mostly depend on the parameters B_2^0 and B_4^0 which may be inferred from a fit to the measured excitation energies. Setting $B_n^m = 0$ for all $m \neq 0$ then yields the results presented in fig. 5.5.

[2]([60], chapter XXXV: "Die Lanthanoide", pg. 1939)

[3]see, e.g., the type II spectrum in fig. 5.2 which is virtually identical to the CuN spectrum and exhibits an excitation-like feature around V=0.

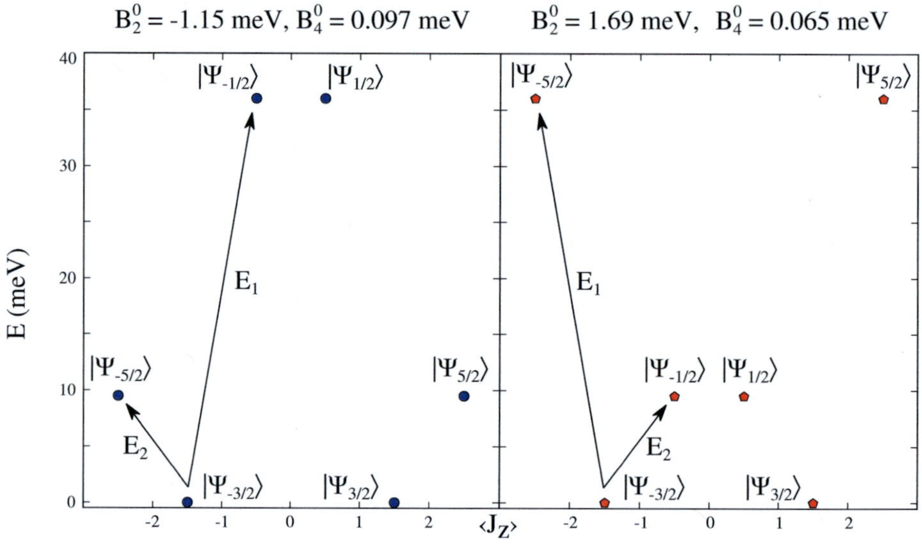

Figure 5.5.: *Schematic of the $\mathcal{H}_{CF}$-spectrum for the two possible scenarios in Ce^{3+} if only two of the measured excitations have a magnetic origin. The crystal field parameters $B^0_{2/4}$ were chosen such that the excitation energies E_1 and E_2 correspond to the indicated transitions.*

In principle, some additional information could be extracted from the spectra in fig. 5.4, such as changes in intensity and width of the dips and peaks with magnetic field, to get an estimate of the crystal field parameters with $m \neq 0$. Unfortunately, this information is hardly usable here: Either the values of a dip-peak pair do not agree very well or their behavior with magnetic field is inconsistent, e.g., one of them becoming narrower while at the same time the other becomes wider, both veering away from their previously matching values. The full set of fit parameters can be found in appendix A.

Adsorption Site

Lastly, the analysis to determine the adsorption site will briefly be presented. As was mentioned earlier, mainly two species of adsorbates were found on the surface, and it was hypothesized that they occupy different adsorption sites. The connection between spectra and adsorption site, however, can only be drawn indirectly since the tip conditions during spectroscopy did not allow atomically resolved imaging and vice versa. Hence, the mapping between the sites and occurrence of spectral features had to be done using the measured height of the adsorbates.

Figure 5.6 shows the determination of the adsorption site of three objects on the surface that are supposed to be cerium adatoms. Two of them match the adsorbates discussed

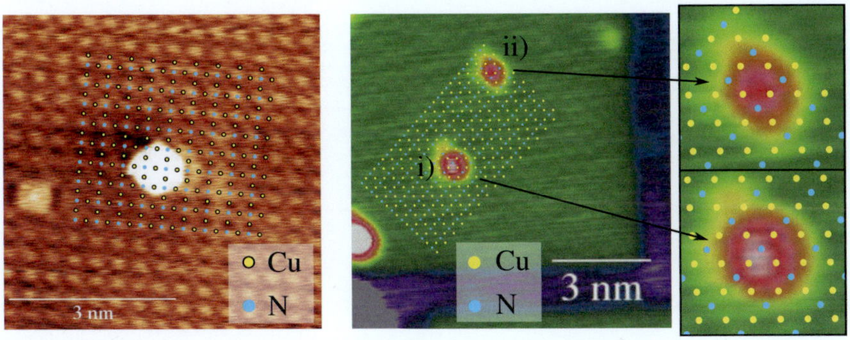

(a) (b)

Figure 5.6.: *Determination of the adsorption site. In atomically resolved images the Cu_2N grid is drawn as an overlay. The adsorption site is then chosen as the lattice point that is closest to the (estimated) geometric center of the adsorbate. While in (a) the bridge position is found for a* ~200 pm *high adsorbate, the positions in (b) are slightly more difficult to determine. The adsorbate i) which is* ~300 pm *high does not match the lattice very well but the highest region tends more towards the top site. The unknown adsorbate ii) above the center (height:* ~260 pm*) seems to occupy a fourfold hollow site.*

above, regarding width and height (see fig. 5.2). The 200 pm high adsorbates (fig. 5.6(a)) are identified with the type I adatoms from above, although no concluding spectra could be taken on this particular adsorbate. The topography images suggest a bridge position ("B") for this adsorbate type. Likewise the ~300 pm adsorbate marked with "i)" in fig. 5.6(b) is assumed to represent type II adatoms, which seem to occupy a top site ("T"). However, the adsorption site determination is less clear in this case, and also for a third adsorbate type labeled "ii)" with intermediate height. This one apparently occupies a fourfold hollow site ("H"), but it was not investigated any further since this type was observed only rarely.

The different adsorption sites may already explain the different apparent height, or they could hint at a different electronic configuration, since adatoms differing in their oxidation state might favor different positions on the surface. Due to the fact that so far the experiments never worked perfectly, a concluding statement about the nature of the different adsorbates cannot be given, yet. Nevertheless it is believed that with the help of supporting calculations and additional measurements focusing on differences in electronic and magnetic behavior, the properties of cerium adatoms on the copper nitride surface can be uncovered.

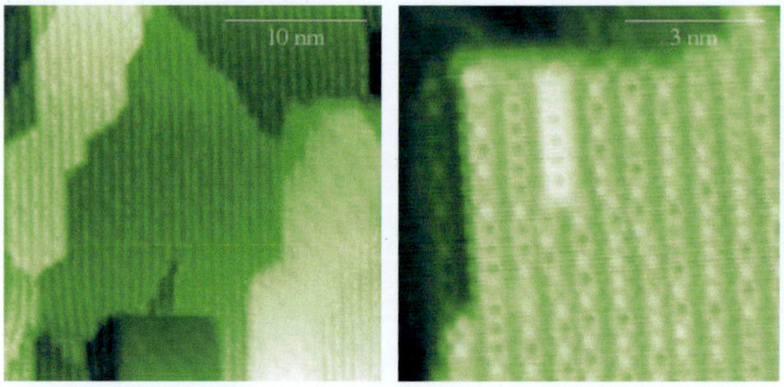

(a) (b) (c)

Figure 5.7.: *Surface of the copper nitride covered Cu(110) crystal. Long narrow stripes form along the [110] direction, partially decorated with additional nitrogen atoms. The close-up (b) shows the irregular occupation of the top sites on a stripe in more detail, as well as an additional short chain above the topmost layer. These occur rarely but they can complicate the proper identification of adsorbate chains vs. Cu_2N structures, especially with less-than optimal tips. (c) shows an overview of the Cu_2N/Cu(110) surface after deposition of Co onto a cold substrate. Several adsorbates of atomic size are visible.*

5.2. Cobalt on Copper Nitride

Although the measurements of magnetic adatoms and clusters on Cu_2N/Cu(100) can be considered quite a breakthrough to reveal spin excitations due to the high signal intensity, the (100)-surface does not support, e.g., self-assembled linear structures. Short antiferromagnetic [53] as well as ferromagnetic chains [61] of transition metal adatoms (Mn and Fe, respectively) on Cu_2N/Cu(100) were studied but had to be built by atomic manipulation. In contrast, a bottom-up approach resulting in one-dimensional structures could possibly be achieved by using anisotropic substrates. Typically, a (110)-surface has different diffusion coefficients along the two main symmetry directions ([001] and [110]) and should thus lead to formation of stripes or chains of adsorbates.[4]

Aside from a general interest in quasi one-dimensional magnetic systems for fundamental research, magnetic nanostructures have become more and more interesting for nano-sized functionality or future spintronics applications. Therefore, studying self-assembling magnetic structures on surfaces similar to the acclaimed Cu_2N/Cu(100) seems like an obvious next step. A few studies with transition metal elements on Cu_2N/Cu(110) were published in the last 15 years [62–64], though the low-energy range (few tens of meV) was never investigated in detail.

In this work, first single cobalt atoms on Cu_2N/Cu(110) were investigated regarding their spin excitations. Then the experiments were extended to self-assembled Co chains, hoping to reveal collective magnetic excitations.

[4]at least as long as they stay on the surface.

Cu_2N on Cu(110)

In contrast to the (100) surface, copper nitride on Cu(110) has received only limited attention from STM studies in the past twenty years. Possibly this is due to its more complex structure compared to that of the Cu(100) surface, showing a strong coverage dependence [65]. Figure 5.7 shows the topography of the clean Cu_2N/Cu(110) surface. On a large scale, flat terraces can be observed covered with long parallel stripes of ~0.8 nm width (see fig. 5.7(a)), constituting a 2×3 structure. On the atomic scale, irregularities of those stripes can be resolved. The basic structure resembles links of a chain in the topography but occasionally the "hole" sites are occupied by a nitrogen atom, shown in fig. 5.7(b). A full coverage of nitrogen leads to all of these sites being occupied. A detailed description can be found, e.g., in [63].

Cobalt Single Atoms

As with almost every deposition, also the behavior of cobalt on copper nitride is sensitive to the substrate temperature. At room temperature and above, diffusion occurs mainly along the [110] direction, leading to chain formation [62, 64]. In order to obtain single atoms from a deposition, the substrate has to be sufficiently cold. The JT-STM does not offer to cool the manipulator, so as a workaround the Cu_2N/Cu(110)-sample was taken from the cold STM and transfered quickly into the preparation chamber where the deposition took place[5]. The temperature of the sample could not be measured, however the resulting adsorbates indicate that no significant diffusion occurred (see fig. 5.7(c)).

Unfortunately, the Co evaporator was never found to be perfectly clean despite repeated degassing cycles so that some of the adsorbates are assumed to be contaminations. This fact also degraded tip conditions since some adsorbates could be picked up easily by the tip. Usually this impaired taking atomically resolved images so that a statement about the adsorption site cannot be made. Furthermore, the identification of Co adatoms was complicated by the fact that almost every adatom showed varying apparent heights between 190 pm and 260 pm. However, due to the size of the surface unit cell and irregularities in the Cu_2N structure, a multitude of different adsorption sites can be expected that might lead to a varying apparent height of otherwise identical adsorbates. Figure 5.8(a) shows several adsorbates that all differ slightly in height and shape.

The results of spectroscopy measurements are presented in fig. 5.8(b). While the adsorbate B in the center and the substrate[6] did not show anything of interest in the spectra the one marked with A exhibits two strong pairs of conductance steps and corresponding dip-peak pairs of high intensity in the d^2I/dV^2 curve.

[5] The evaporator at the STM chamber that points directly at the sample was occupied at that time with a different material.

[6] or, in fact, any other adsorbate in the image

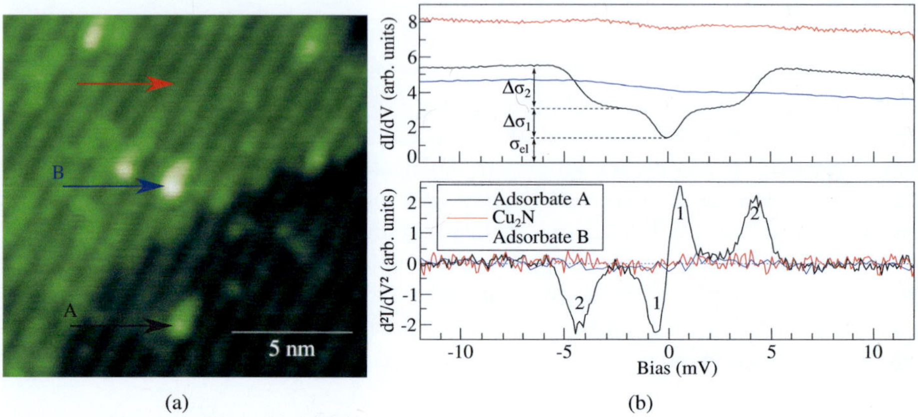

Figure 5.8.: *Close-up (a) of adsorbates that were studied with spectroscopy techniques. The short, blurry "tail" at each adsorbate is merely due to the tip. First and second derivative spectra (b) were taken on copper nitride (red arrow) as well as on two different adsorbates (black, blue). Except for adsorbate* A *none of the adsorbates exhibited any excitation features in the spectra. Experimental parameters were* $T \approx 880\,\mathrm{mK}$, $I = 1\,\mathrm{nA}$, $V = 12\,\mathrm{mV}$, $V_{\mathrm{mod}} = 150\,\mu\mathrm{V}$, *B=0.*

The excitation with higher energy could easily be resolved, the lower-energy excitation signatures, in contrast, overlap at zero bias and make it difficult to reliably determine the excitation energy, intensity, and level broadening. By fitting Gaussians to the spectra the excitation energies were found to be $E_1 = 0.4 \pm 0.1\,\mathrm{meV}$ and $E_2 = 4.26 \pm 0.02\,\mathrm{meV}$, however, since the energy resolution at $T = 880\,\mathrm{mK}$ and $V_{\mathrm{mod}} = 150\,\mu\mathrm{eV}$ is approximately 0.5 meV and thus larger than E_1 this energy can only be considered an upper limit for that particular excitation. Similarly, an estimate for the excitation cross section suffers from a reduced accuracy since the zero-bias conductance cannot be determined unambiguously due to the limited energy resolution. Nevertheless, using the measured value yields that excitation 1 has a probability roughly equal to the elastic process, i.e., the conductance rises by $\sim 100\,\%$, whereas excitation 2 leads to an even higher increase of $\sim 160\,\%$. Again, both values have to be considered as lower bounds.

Within the limits of what the JT-STM could allow, the dependence of the excitations on an external *Bz* field was studied. Spectra were taken at 1 T and 1.5 T and the excitation energies were determined in the same way as before. The result of the fitting procedures is presented in fig. 5.9 where the energies of both excitations are plotted. While the lower energy excitation does not show a clear Zeeman shift, the higher one does. Surprisingly, $E_2(1\mathrm{T})$ is found to be $\sim 450\,\mu\mathrm{eV}$ smaller than without field, but at 1.5 T it is again approaching its zero-field value. Quantitatively, however, the shift of E_2 is quite large compared to the value of $57.9\,\mu\mathrm{eV\,T^{-1}}$ for one μ_B and would imply a very large magnetic moment.

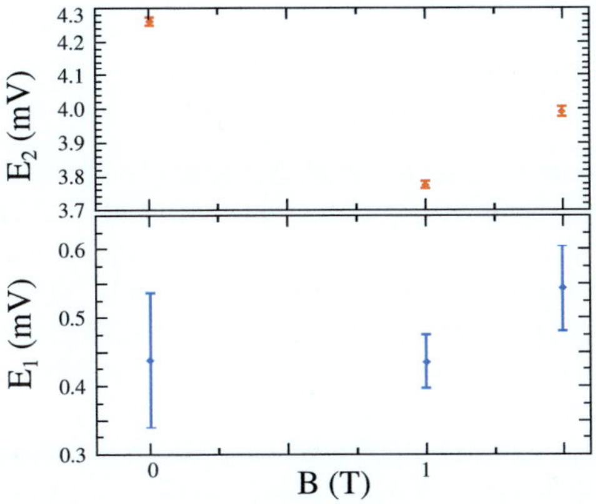

Figure 5.9.: *Observed Zeeman shift of the excitation energies $E_{1/2}$. The lower energy shift is not significant, the shift of E_2 however is surprisingly strong at* 1 T *and is then reduced at higher field. Experimental conditions were as in fig. 5.8(b) except for T. The temperature was* 915 mK *at* 1 T *and* 965 mK *at* 1.5 T*, though there is no guarantee that the temperature sensor is immune to the magnetic field.*

The interpretation in terms of eigenstates of a crystal field Hamiltonian is difficult, though: On the one hand, since the rotational symmetry of the (110) surface is lower than in the (100), case the symmetry of the adsorption site can safely be assumed as C_{2v} or less. An accidentally occurring fourfold symmetric site seems highly improbable. On the other hand, for the particular system of Co adatoms on Cu_2N/Cu(110) no further information is available about the magnetic properties, such as spin and orbital angular momentum. The similar system Co on Cu_2N/Cu(100) is assumed to have $S = 3/2$ [39][7] but this is not necessarily true for the (110) surface. Furthermore, some non-zero orbital moment must exist since otherwise no magnetic anisotropy could emerge from the crystal field. Lastly, it is impossible to obtain two excitations for a $J = 3/2$ system with $\vec{B} = 0$, which means that either the Co adatom must have a different, presumably integer, spin or total angular momentum, or one of the two excitations is not of magnetic origin. Therefore, a conclusive statement cannot be given at this stage – more experiments need to be conducted in order to clarify the field dependence of the excitations and to find out the adsorption site(s).

Cobalt Chains

In a 2014 study Fe adatoms on Cu_2N/Cu(100) were assembled into chains of five and six atoms and their excitation spectrum was resolved depending on the position along the chain [61]. In contrast to single atom spectroscopy the excitations in these chains are the

[7] and exhibits a Kondo resonance depending on the position of the adatom with respect to the boundaries of the Cu_2N-islands [35] which was not seen at all in the experiments presented here.

elementary collective excitations, i.e., magnons, that emerge due to exchange coupling between the atoms. An investigation comparable to that study was one of the goals for the experiments on the $Cu_2N/Cu(110)$ surface in this thesis.

Hence, after finding spectroscopic features that strongly resemble magnetic excitations in single adatoms, self-assembled Co-chains were studied on this surface. For this, the deposition procedure was changed: more material (though still less than a monolayer) was deposited while the substrate temperature was roughly 50 °C to 80 °C. An example of the resulting topography is shown in fig. 5.10, where three cobalt chains of 5 to 10 nm length can be seen. This is qualitatively similar to the previously reported growth of Co and other transition metals on this surface [62–64]. The Co coverage is obviously rather low so that these chains can be considered isolated, i.e., presumably not interacting. A detailed model for the atomic arrangement in similar Fe chains is described in [63]. Essentially, each chain is made up of five parallel one-atomic rows, but not all atomic adsorption sites are equivalent. If this arrangement is assumed for Co chains as well, a multitude of discrete magnon modes must be expected that could be difficult to resolve spectroscopically when taking finite magnon lifetimes into account. However, it was not possible during the repeated measurements to obtain atomically resolved topography images of these chains so that a clear confirmation of the atomic structure could not yet be achieved.

Moreover, the measurements could not repeat the clear findings from the single atom case. Chains of different lengths – from 3 nm to 10 nm – were investigated with first and second derivative spectroscopy. However, no excitations could be found, in fact, the spectra taken on Co chains did not reveal any systematic deviation from those obtained on the Cu_2N substrate, neither at low bias voltages nor up to 200 mV. Therefore, possible follow-up investigations with varying magnetic field were not performed.

Figure 5.10.: *Formation of Co chains on the $Cu_2N/Cu(110)$ surface after deposition at elevated temperatures. The chains are approximately* 1 nm *wide and several nanometers long; at higher Co dosage, longer chains and an increased chain density could be observed.*

Among the hypothetical explanations for this behavior one could name the already mentioned fact that the cleanliness of the Co evaporator was not optimal so that the observed chains may have contained too many impurities. Another reason, however, could be a similar finding for monatomic chains on Cu_2N/Cu(100) [66]: There, excitations could only be observed in the outermost atoms; all other atoms in the chain did not show any typical signatures. This was explained by a geometric rearrangement and a stronger incorporation of the Co atoms into the Cu_2N structure, leading to an altered filling of the d-orbitals that effectively inhibits spin excitations and partially suppresses orbital excitations by tunneling electrons. Whether this happens also on the Cu_2N/Cu(110) surface could not yet be investigated, but this scenario could certainly motivate some clarifying calculations or renewed efforts to atomically resolve the structure of the chains.

Chapter 6.

6 Conclusion

The present thesis focused on the investigation of various magnetic properties of single adatoms. One of the fundamental motivations behind this research area is the desire to, one day, use single magnetic atoms as memory. In contrast to other single atom studies before, the system of holmium atoms on Pt(111) seemed like a promising candidate to uncover the principles that either allow or prevent stable spins. Thus, the goal was to continue the investigation of this system in order to confirm and advance the results of surprisingly long lifetimes of the atom's spin, previously reported by Schuh et al.

For this purpose, scanning tunneling microscopy and the related spectroscopy capabilities were used to study spin excitations and, more importantly, the dynamics of the magnetic ground states by means of spin-polarized STM. In the focus of the experiments was the investigation of the lifetime dependencies on various parameters, first and foremost of an external magnetic field, but also tunneling voltage, temperature, and tunneling current. Although the experiments turned out to be extremely challenging due to the usage of difficult-to-handle antiferromagnetic tips and partially operating the STM close to its performance limits, finally a sufficiently convincing series of experiments could be conducted. The most important result is the confirmation of magnetic switching between two states with long lifetimes, unprecedented in single atoms on metallic surfaces.

Since the magnetic properties of holmium atoms originate from the $4f$ shell, a theoretical description with a crystal field Hamiltonian, expressed in terms of Stevens operators is appropriate. This is in contrast to the commonly studied transition metal adatoms where the orbital momentum of the $3d$ shell is usually quenched and does not contribute to the magnetic properties. The investigation of the Hamiltonian's spectrum and the subsequent comparison with experimental results was made possible by ab-initio calculations from the collaborating group at MPI Halle. Especially, these calculations justified the empirically motivated assumption that Ho adatoms on Pt(111) are in their 3+ oxidation state, establishing a $^5I_8$ ground state multiplet according to Hund's rules. Thus it could be confirmed that the very fortunate combination of a threefold symmetric adsorption site with the properties of the $4f$ shell is realized in this system. The two fundamental ingredients for the observed stability are the formation of two degenerate ground states by the crystal field and their protection from single-electron induced magnetization reversal due to time-inversion symmetry.

Furthermore, the ab-initio calculations enabled the evaluation of a quantitative description with two master equation approaches — one classical, one quantum — by providing the anisotropy constants. These efforts faced substantial difficulties that are not yet completely resolved due to the many available parameters but only a comparably scarce amount of experimental results. Nevertheless, with the newer quantum master equation calculations, it is believed that an appropriate quantitative description can be obtained, which will also help to identify the most lucrative measurements in order to determine more and more of the model parameters. Meanwhile, new experiments to complement the existing results are planned and will hopefully be carried out in the near future.

Motivated by the Ho on Pt(111) results, a prediction for the other $4f$ elements was made, identifying several other highly interesting combinations of adatom and adsorption site symmetry. Among them, the particularly interesting systems of adatoms with $J = 5/2$ in sixfold symmetric sites should be thoroughly investigated.

Two other projects on magnetism of single adatoms were started during this work: Inspired by the holmium findings, cerium adatoms on the widely-used Cu_2N/Cu(100) surface were investigated, revealing two species of adatoms with different apparent height and different excitation spectra. Although a final statement about the origin of the differences cannot be given, yet, it is speculated that the oxidation state or the adsorption site — or a combination of both — may play an important role. Further investigations are strongly encouraged.

The findings on cobalt single atoms on the Cu_2N/Cu(110) surface also require further experiments. There as well, the origin of the observed excitation spectrum is not yet fully understood. A proper determination of the adsorption site, as well as conclusive experiments studying the magnetic field dependence of the excitations would be the high-priority goals when reinvestigating this system. The discovery of apparently non-magnetic self-assembled cobalt chains on this surface does not immediately call for a renewed investigation. Of course, the experimental determination of the atomic structure, supported by calculations, would certainly help to explain the findings. However, following the initial motivation to explicitly study excitations in self-assembled magnetic chains would suggest finding other combinations of adsorbate and substrate that permit the investigation of such structures.

Lastly, one part of this work was dedicated to the task of completing the research group's newest piece of experimental equipment (the "d-STM") by designing and manufacturing a custom superconducting coil. The incentive lies in the desire to enable (among others) all the above-mentioned experiments that need to apply high magnetic fields in order to study spin excitations. Although the coil did not achieve its full theoretical potential during the two test runs, it can be considered a successful construction, yielding many valuable lessons for future projects. With a proper safety margin, out-of plane oriented magnetic fields of up to 6 T should be applicable in the new setup.

In conclusion, it is probably fair to state that the understanding of the holmium case has been significantly advanced, although many open questions still remain. With the two additional adatom studies just begun and a newer, better STM available with lower temperatures and higher fields, many interesting results on single magnetic adatoms can be expected in the future.

Acknowledgements

The work that was done on these projects in the last three and a half years would not have been possible for me without many people. I would like to thank

- Prof. Dr. Wulf Wulfhekel for the possibility to work on this subject and his confidence in me, for his support and seemingly endless optimism, for ideas, explanations, and his positive attitude towards all of his group members;
- Prof. Dr. Georg Weiß for being the second reviewer of this work;
- Prof. Dr. Arthur Ernst and his team for the *ab initio* calculations for several of our single atom systems;
- our collaborators from TFP, namely Prof. Dr. Gerd Schön, Dr. Michael Marthaler, and Christian Karlewski for their (quantum) master equation calculations and fruitful discussions, as well as proof-reading of my thesis (C. K.);
- Dr. Toshio Miyamachi for being an incredible coworker, his efforts with our experiments, his countless night shifts during which he managed to measure things nobody else could measure, and for the great times we had;
- Dr. Timofey Balashov for his willingness to answer innumerable questions about single adatoms and physics in general, for support, advice and feedback, for proof-reading, computer magic and crazy conversations;
- Dr. Tobias Schuh, my predecessor, for the great work he did that let me take over a very interesting and ambitious project; and Alexander Stöhr, a former colleague and Diploma student who worked also with T. S. on this project;
- Christopher Bresch for his tremendous work on our holmium project;
- Martin Obermair who worked on the measurements on cerium and cobalt single atoms with me and Clemens Baretzky who will eagerly continue this research, as well as Dr. Hironari Isshiki for his interest and recurring support during the experiments;
- Dominik Rau who built the superconducting coil with me;
- all group members — former and present — of AG Wulfhekel, especially Marie, Lukas, Lei, Michael, Moritz, Jasmin and Jinjie for creating the great atmosphere at work;

- Christina for proof-reading and the occasional cheering-up;
- my family and friends for their support, love, friendship, cake, and companionship!

A

Appendix A.

Appendix

Hund's Rules

These rules are a set of empirical results or instructions to describe how to determine the total angular momentum characteristics of the ground state multiplet of a multi-electron system with a partially filled subshell. They can be found in standard textbooks on quantum mechanics or atomic physics in several different formulations.

- Completely filled subshells have $L = 0$ and $S = 0$ and thus do not contribute to J.
- The lowest multiplet has maximum S.
- Of the remaining possible constellations, the one with maximum L has the lowest energy.
- Finally, these two quantities combine to form J: for less than half-filled subshells, $J = |L - S|$, otherwise $J = L + S$.

Anisotropy and Adsorption Site Symmetry

Stevens Operators

This is a selection of Stevens operators O_n^m that are in some way relevant for this thesis. Wherever $m \neq 0$, the $\tilde{O}_n^m$ can be obtained by changing the relative sign in the $J_+^m + J_-^m$ term, and by introducing a global factor of $-i$. $\{A, B\}$ is a shorthand notation for the anticommutator $AB + BA$.

Second order terms:

$$
\begin{aligned}
O_2^0 &= 3J_z^2 - \vec{J}^2 \\
O_2^1 &= \frac{1}{4}\{J_z, (J_+ + J_-)\} \\
O_2^2 &= \frac{1}{2}\left[J_+^2 + J_-^2\right]
\end{aligned}
\tag{A.1}
$$

Fourth order terms:

$$
\begin{aligned}
O_4^0 &= 35J_z^4 - 30\vec{J}^2 J_z^2 + 25J_z^2 - 6\vec{J}^2 + 3(\vec{J}^2)^2 \\
O_4^1 &= \frac{1}{4}\left\{\left(7J_z^3 - (3\vec{J}^2 + 1)J_z\right), (J_+ + J_-)\right\} \\
O_4^2 &= \frac{1}{4}\left\{\left(7J_z^2 - \vec{J}^2 - 5\right), (J_+^2 + J_-^2)\right\} \\
O_4^3 &= \frac{1}{4}\{J_z, (J_+^3 + J_-^3)\} \\
O_4^4 &= \frac{1}{2}\left[J_+^4 + J_-^4\right]
\end{aligned}
\tag{A.2}
$$

Sixth order terms:

$$
\begin{aligned}
O_6^0 &= 231J_z^6 - 315\vec{J}^2 J_z^4 + 735J_z^4 + 105(\vec{J}^2)^2 J_z^2 \\
&\quad - 525\vec{J}^2 J_z^2 + 294J_z^2 - 5(\vec{J}^2)^3 + 40(\vec{J}^2)^2 - 60\vec{J}^2 \\
O_6^1 &= \frac{1}{4}\left\{\left(33J_z^5 - (30\vec{J}^2 - 15)J_z^3 + (5\vec{J}^2 - 10\vec{J}^2 + 12)J_z\right), (J_+ + J_-)\right\} \\
O_6^2 &= \frac{1}{4}\left\{\left(33J_z^4 - (18\vec{J}^2 + 123)J_z^2 + (\vec{J}^2)^2 + 10\vec{J}^2 + 102\right), (J_+^2 + J_-^2)\right\} \\
O_6^3 &= \frac{1}{4}\left\{\left(11J_z^3 - 3\vec{J}^2 J_z - 59J_z\right), (J_+^3 + J_-^3)\right\} \\
O_6^4 &= \frac{1}{4}\left\{\left(11J_z^2 - \vec{J}^2 - 38\right), (J_+^4 + J_-^4)\right\} \\
O_6^5 &= \frac{1}{4}\{J_z, (J_+^5 + J_-^5)\} \\
O_6^6 &= \frac{1}{2}\left[J_+^6 + J_-^6\right]
\end{aligned}
\tag{A.3}
$$

Nonzero Coefficients

symm.	B_2^0	B_2^1	B_2^2	B_4^0	B_4^1	B_4^2	B_4^3	B_4^4	B_6^0	B_6^1	B_6^2	B_6^3	B_6^4	B_6^5	B_6^6
C_1	+	•	•	+	•	•	•	•	+	•	•	•	•	•	•
$C_{1v}(C_s)$	+	∘	∘	+	∘	∘	∘	∘	+	∘	∘	∘	∘	∘	∘
C_2	+		•	+		•		•	+		•		•		•
C_{2v}	+		∘	+		∘		∘	+		∘		∘		∘
C_3	+			+			•		+			•			•
C_{3v}	+			+			∘		+			∘			∘
C_4	+			+				•	+				•		
C_{4v}	+			+				∘	+				∘		
C_6	+			+					+						•
C_{6v}	+			+					+						∘

Table A.1.: *This table lists the coefficients B_n^m for the corresponding operators that can appear in a given point symmetry on a surface. The symbol • means that both B and $\tilde{B}$ are nonzero, ∘ indicates that one of them can vanish (through choice of coordinate system). Finally, + signifies that the corresponding operator will always be present, given that the constraints of $n \leqslant 2J$ or $n \leqslant 2\ell$ allow it, of course.*

Correct Lifetime Estimation from ITS Curves

In a simple description, the total width of a peak (dip) in an ITS curve can be accounted for by approximating the respective broadening due to modulation voltage, temperature, and finite lifetimes with Gaussian curves, as was done in eq. (2.8). When V_{mod} and T are known, the intrinsic linewidth Γ (FWHM) can be used to calculate the lifetime τ via the uncertainty principle:

$$\tau = \frac{\hbar}{\Gamma} \tag{A.4}$$

If a more realistic approach is used, the line shape of the excitation must be described by a Lorentzian curve. Furthermore, the actual distributions ϕ and χ due to modulation and temperature must be used [19]. Previously, the computationally intensive convolution of the three functions could be avoided; now it needs to be evaluated when fitting the resulting curve to a ITS curve, though. Typically, this gives a narrower linewidth for the Lorentzian than the simpler Gaussian approach, and thus the estimated lifetimes are longer.

With $\beta = 1/k_\mathrm{B}T$ and $x = eV$, the used functions read:

$$f(x, x_0, \Gamma) = \frac{1}{(x - x_0)^2 + \Gamma^2/4} \tag{A.5}$$

$$\phi(x) = \beta e^{\beta x} \frac{(\beta x - 2)e^{\beta x} + \beta x + 2}{(e^{\beta x} - 1)^3} \tag{A.6}$$

$$\chi(x) = \begin{cases} (2(eV_\mathrm{mod})^2 - x^2)^{3/2} & , |x| < \sqrt{2}eV_\mathrm{mod} \\ 0 & , |x| \geqslant \sqrt{2}eV_\mathrm{mod} \end{cases} . \tag{A.7}$$

The full fit curve then becomes

$$d(x) = \sum_i A_i (f(x_i, \Gamma_i) * \phi * \chi)(x), \tag{A.8}$$

where $*$ denotes a convolution and A_i is a fitting amplitude for the ith peak or dip which is centered at x_i. If the spectra allow, one can also attempt to fit a dip-peak pair with a common set of parameters ($\pm x_i, \pm A_i, \Gamma_i$). Additionally, a linear background term may sometimes be necessary to account for a tilt and vertical offset of the spectrum, but the resulting fit parameters do not yield any physical quantities and are thus omitted in the above discussion.

Fit Parameters for Ce on Cu_2N/Cu(100)

Here, the full set of fit parameters from the spectroscopy experiments with magnetic field as described in section 5.1 will be presented. Instead of using the elaborate procedure above, for simplicity six normalized Gaussians were fitted to the measured spectra:

$$\frac{d^2I}{dV^2}(V) = \sqrt{\frac{2}{\pi}} \sum_{i=1}^{6} \frac{A_i}{\Gamma_i} \exp[-2(eV - E_i)^2/\Gamma_i^2] \tag{A.9}$$

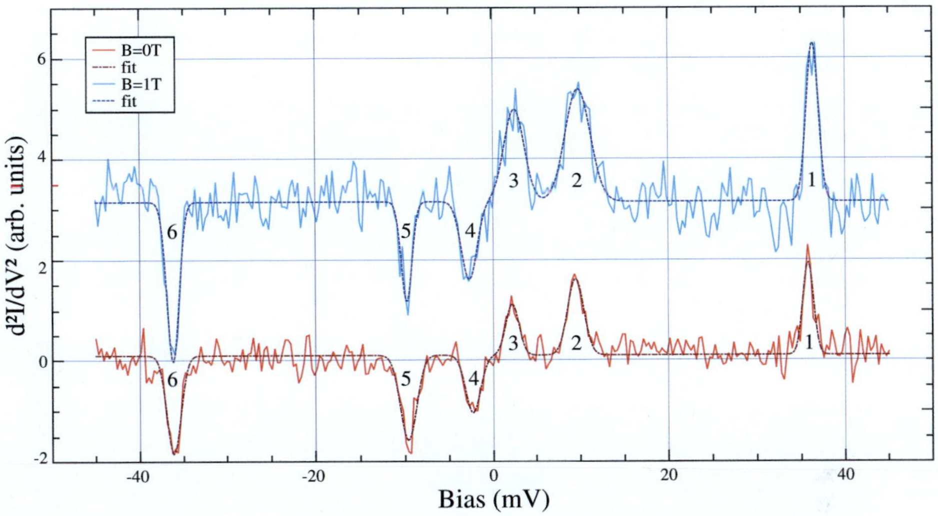

Figure A.1.: *Enlarged version of the spectra in fig. 5.4, with additional labels (1-6) to denote all features in the recorded curves. In section 5.1 the fitted values for the pairs 1-6, 2-5 and 3-4 were averaged.*

	B=0T						B=1T					
	E	σ_E	A	σ_A	Γ	σ_Γ	E	σ_E	A	σ_A	Γ	σ_Γ
(1)	35.73	0.05	28.2	0.2	1.2	0.1	36.28	0.05	54.1	0.4	1.4	0.1
(2)	9.41	0.08	34.4	0.3	1.8	0.2	9.7	0.1	79.1	0.6	2.8	0.2
(3)	2.2	0.1	19.0	0.2	1.7	0.2	2.5	0.1	53.2	0.5	2.3	0.2
(4)	-2.3	0.1	-24	0.3	1.5	0.2	-2.7	0.1	-34.0	0.4	1.8	0.3
(5)	-9.60	0.07	-38.7	0.3	1.8	0.1	-9.74	0.08	-30.1	0.4	1.2	0.2
(6)	-36.14	0.06	-39.5	0.3	1.6	0.1	-36.21	0.05	-55.8	0.4	1.4	0.1

Table A.2.: *Fit parameters of six Gaussians fitted to the data in fig. A.1. Energies E and widths Γ are given in meV; intensities A have arbitrary units (but are scaled by 1000) since only the relative values are important.*

Bibliography

[1] R. Wood, Mason Williams, A. Kavcic, and Jim Miles. The Feasibility of Magnetic Recording at 10 Terabits Per Square Inch on Conventional Media. *Magnetics, IEEE Transactions on*, 45(2):917–923, 2009. URL `10.1109/TMAG.2008.2010676`.

[2] M.H. Kryder, E.C. Gage, T.W. McDaniel, W.A. Challener, R.E. Rottmayer, Ganping Ju, Yiao-Tee Hsia, and M.F. Erden. Heat Assisted Magnetic Recording. *Proceedings of the IEEE*, 96(11):1810–1835, 2008. URL `10.1109/JPROC.2008.2004315`.

[3] Masamitsu Hayashi, Luc Thomas, Rai Moriya, Charles Rettner, and Stuart S. P. Parkin. Current-Controlled Magnetic Domain-Wall Nanowire Shift Register. *Science*, 320(5873):209–211, 2008. URL `http://www.sciencemag.org/content/320/5873/209.abstract`.

[4] V. Madhavan, W. Chen, T. Jamneala, M. F. Crommie, and N. S. Wingreen. Tunneling into a Single Magnetic Atom: Spectroscopic Evidence of the Kondo Resonance. *Science*, 280(5363):567–569, 1998. URL `http://dx.doi.org/10.1126/science.280.5363.567`.

[5] B. C. Stipe, M. A. Rezaei, and W. Ho. Single-Molecule Vibrational Spectroscopy and Microscopy. *Science*, 280(5370):1732–1735, 1998. URL `http://dx.doi.org/10.1126/science.280.5370.1732`.

[6] A. J. Heinrich, J. A. Gupta, C. P. Lutz, and D. M. Eigler. Single-Atom Spin-Flip Spectroscopy. *Science*, 306(5695):466–469, 2004. URL `http://dx.doi.org/10.1126/science.1101077`.

[7] P. Gambardella, S. Rusponi, M. Veronese, S. S. Dhesi, C. Grazioli, A. Dallmeyer, I. Cabria, R. Zeller, P. H. Dederichs, K. Kern, C. Carbone, and H. Brune. Giant Magnetic Anisotropy of Single Cobalt Atoms and Nanoparticles. *Science*, 300(5622):1130–1133, 2003. URL `http://dx.doi.org/10.1126/science.1082857`.

[8] Sebastian Loth, Susanne Baumann, Christopher P. Lutz, D. M. Eigler, and Andreas J. Heinrich. Bistability in Atomic-Scale Antiferromagnets. *Science*, 335(6065):196–199, 2012. URL `http://dx.doi.org/10.1126/science.1214131`.

[9] Toshio Miyamachi, Tobias Schuh, Tobias Märkl, Christopher Bresch, Timofey Balashov, Alexander Stöhr, Christian Karlewski, Stephan Andre, Michael Marthaler, Martin Hoffmann, Matthias Geilhufe, Sergey Ostanin, Wolfram Hergert, Ingrid Mertig, Gerd Schön, Arthur Ernst, and Wulf Wulfhekel. Stabilizing the magnetic moment of single holmium atoms by symmetry. *Nature*, 503(7475):242–246, 2013. URL `http://dx.doi.org/10.1038/nature12759`.

[10] Tobias Schuh. *Spin-Dynamik in Atomen und Clustern auf Oberflächen*. PhD thesis, KIT, 2012. URL `http://digbib.ubka.uni-karlsruhe.de/volltexte/1000028761`. Karlsruhe, KIT, Diss., 2012.

[11] G. Binnig, H. Rohrer, Ch. Gerber, and E. Weibel. Surface Studies by Scanning Tunneling Microscopy. *Phys. Rev. Lett.*, 49:57–61, 1982. URL `http://dx.doi.org/10.1103/PhysRevLett.49.57`.

[12] Edward L. Wolf. *Principles of electron tunneling spectroscopy*. International series of monographs on physics; 71. Oxford Univ. Pr., New York, Paperback edition, 1989.

[13] Chunli Bai. *Scanning tunneling microscopy and its applications*. Springer series in surface sciences ; 32. Springer, Berlin, 2nd, rev. edition, 2000.

[14] C. Julian Chen. *Introduction to scanning tunneling microscopy*. Monographs on the physics and chemistry of materials ; 64. Oxford University Press, Oxford, 2nd edition, 2008.

[15] J. Bardeen. Tunnelling from a Many-Particle Point of View. *Phys. Rev. Lett.*, 6: 57–59, 1961. URL `http://dx.doi.org/10.1103/PhysRevLett.6.57`.

[16] J. Tersoff and D. R. Hamann. Theory of the scanning tunneling microscope. *Phys. Rev. B*, 31:805–813, 1985. URL `http://dx.doi.org/10.1103/PhysRevB.31.805`.

[17] I. Giaever, H. R. Hart, and K. Megerle. Tunneling into Superconductors at Temperatures below 1° K. *Phys. Rev.*, 126:941–948, 1962. URL `http://link.aps.org/doi/10.1103/PhysRev.126.941`.

[18] J. Lambe and R. C. Jaklevic. Molecular vibration spectra by inelastic electron tunneling. *Phys. Rev.*, 165:821–832, Jan 1968. URL `http://link.aps.org/doi/10.1103/PhysRev.165.821`.

[19] J. Klein, A. Léger, M. Belin, D. Défourneau, and M. J. L. Sangster. Inelastic-Electron-Tunneling Spectroscopy of Metal-Insulator-Metal Junctions. *Phys. Rev. B*, 7:2336–2348, 1973. URL `http://dx.doi.org/10.1103/PhysRevB.7.2336`.

[20] M. Julliere. Tunneling between ferromagnetic films. *Physics Letters A*, 54(3):225 – 226, 1975. URL http://www.sciencedirect.com/science/article/pii/0375960175901747.

[21] D. Wortmann, S. Heinze, Ph. Kurz, G. Bihlmayer, and S. Blügel. Resolving complex atomic-scale spin structures by spin-polarized scanning tunneling microscopy. *Phys. Rev. Lett.*, 86:4132–4135, Apr 2001. URL http://link.aps.org/doi/10.1103/PhysRevLett.86.4132.

[22] L. Zhang, T. Miyamachi, T. Tomani, R. Dehm, and W. Wulfhekel. A compact sub-Kelvin ultrahigh vacuum scanning tunneling microscope with high energy resolution and high stability. *Review of Scientific Instruments*, 82:103720, 2011. URL http://dx.doi.org/10.1063/1.3646468.

[23] Lei Zhang. *Sub-Kelvin Scanning Tunneling Microscopy on Magnetic Molecules*. PhD thesis, KIT, 2012. URL http://digbib.ubka.uni-karlsruhe.de/volltexte/1000030706. Karlsruhe, KIT, Diss., 2012.

[24] A. Kubetzka, M. Bode, O. Pietzsch, and R. Wiesendanger. Spin-Polarized Scanning Tunneling Microscopy with Antiferromagnetic Probe Tips. *Phys. Rev. Lett.*, 88: 057201, 2002. URL http://link.aps.org/doi/10.1103/PhysRevLett.88.057201.

[25] Yukikazu Iwasa. *Case Studies in Superconducting Magnets : Design and Operational Issues*. Springer, New York, 2nd edition, 2009.

[26] Martin N. Wilson. *Superconducting Magnets*. Monographs on cryogenics ; 2. Clarendon Press, Oxford, 1983.

[27] Supercon homepage. URL http://www.supercon-wire.com/content/nbti-superconducting-wires.

[28] Inc. Wolfram Research. Mathematica, 2012. URL http://www.wolfram.com/mathematica/. Version 9.0.

[29] D. C. Meeker. Finite Element Method Magnetics (FEMM), April 2012. URL http://www.femm.info. Version 4.2.

[30] Tobias Schuh, Timofey Balashov, Toshio Miyamachi, Shih-Yu Wu, Chein-Cheng Kuo, Arthur Ernst, Jürgen Henk, and Wulf Wulfhekel. Magnetic anisotropy and magnetic excitations in supported atoms. *Phys. Rev. B*, 84(10):104401, 2011. URL http://dx.doi.org/10.1103/PhysRevB.84.104401.

[31] R. J. Elliott and K. W. H. Stevens. The theory of magnetic resonance experiments on salts of the rare earths. *Proceedings of the Royal Society of London A: Mathematical, Physical and Engineering Sciences*, 218(1135):553–566, 1953.

[32] Anatole Abragam and Brebis Bleaney. *Electron paramagnetic resonance of transition ions*. The international series of monographs on physics. Clarendon Press, Oxford, 1970.

[33] F. Delgado and J. Fernández-Rossier. Storage of Classical Information in Quantum Spins. *Phys. Rev. Lett.*, 108:196602, 2012. URL http://link.aps.org/doi/10.1103/PhysRevLett.108.196602.

[34] J. Fransson. Spin Inelastic Electron Tunneling Spectroscopy on Local Spin Adsorbed on Surface. *Nano Letters*, 9(6):2414–2417, 2009. URL http://dx.doi.org/10.1021/nl901066a.

[35] Jenny C. Oberg, M. Reyes Calvo, Fernando Delgado, Maria Moro-Lagares, David Serrate, David Jacob, Joaquin Fernandez-Rossier, and Cyrus F. Hirjibehedin. Control of single-spin magnetic anisotropy by exchange coupling. *Nat. Nano.*, 9(1): 64–68, 2014. URL http://dx.doi.org/10.1038/nnano.2013.264.

[36] Christian Karlewski, Michael Marthaler, Tobias Märkl, Timofey Balashov, Wulf Wulfhekel, and Gerd Schön. Magnetic adatoms as memory bits: A quantum master equation analysis. *Phys. Rev. B*, 91:245430, 2015. URL http://link.aps.org/doi/10.1103/PhysRevB.91.245430.

[37] Jiutao Li, Wolf-Dieter Schneider, Richard Berndt, and Bernard Delley. Kondo scattering observed at a single magnetic impurity. *Phys. Rev. Lett.*, 80:2893–2896, 1998. URL http://link.aps.org/doi/10.1103/PhysRevLett.80.2893.

[38] Cyrus F. Hirjibehedin, Chiung-Yuan Lin, Alexander F. Otte, Markus Ternes, Christopher P. Lutz, Barbara A. Jones, and Andreas J. Heinrich. Large Magnetic Anisotropy of a Single Atomic Spin Embedded in a Surface Molecular Network. *Science*, 317(5842):1199–1203, 2007. URL http://dx.doi.org/10.1126/science.1146110.

[39] Alexander F. Otte, Markus Ternes, Kirsten von Bergmann, Sebastian Loth, Harald Brune, Christopher P. Lutz, Cyrus F. Hirjibehedin, and Andreas J. Heinrich. The role of magnetic anisotropy in the Kondo effect. *Nat. Phys.*, 4(11):847–850, 2008. URL http://dx.doi.org/10.1038/nphys1072.

[40] T. Schuh, T. Balashov, T. Miyamachi, A. F. Takcs, S. Suga, and W. Wulfhekel. Lifetimes of magnetic excitations in fe and co atoms and clusters on pt(111). *Journal of Applied Physics*, 107(9), 2010. URL http://scitation.aip.org/content/aip/journal/jap/107/9/10.1063/1.3365113.

[41] Tobias Schuh, Toshio Miyamachi, Stefan Gerstl, Matthias Geilhufe, Martin Hoffmann, Sergey Ostanin, Wolfram Hergert, Arthur Ernst, and Wulf Wulfhekel. Magnetic Excitations of Rare Earth Atoms and Clusters on Metallic Surfaces. *Nano Letters*, 12(9):4805–4809, 2012. URL http://dx.doi.org/10.1021/nl302250n.

[42] Focko Meier, Lihui Zhou, Jens Wiebe, and Roland Wiesendanger. Revealing Magnetic Interactions from Single-Atom Magnetization Curves. *Science*, 320(5872): 82–86, 2008. URL http://www.sciencemag.org/content/320/5872/82.abstract.

[43] Alexander Ako Khajetoorians, Benjamin Baxevanis, Christoph Hübner, Tobias Schlenk, Stefan Krause, Tim Oliver Wehling, Samir Lounis, Alexander Lichtenstein, Daniela Pfannkuche, Jens Wiebe, and Roland Wiesendanger. Current-Driven Spin Dynamics of Artificially Constructed Quantum Magnets. *Science*, 339(6115): 55–59, 2013. URL http://dx.doi.org/10.1126/science.1228519.

[44] Sebastian Loth, Kirsten von Bergmann, Markus Ternes, Alexander F. Otte, Christopher P. Lutz, and Andreas J. Heinrich. Controlling the state of quantum spins with electric currents. *Nat. Phys.*, 6(5):340–344, 2010. URL http://dx.doi.org/10.1038/nphys1616.

[45] C. Hübner, B. Baxevanis, A. A. Khajetoorians, and D. Pfannkuche. Symmetry effects on the spin switching of adatoms. *Phys. Rev. B*, 90:155134, 2014. URL http://link.aps.org/doi/10.1103/PhysRevB.90.155134.

[46] M. A. Ruderman and C. Kittel. Indirect Exchange Coupling of Nuclear Magnetic Moments by Conduction Electrons. *Phys. Rev.*, 96:99–102, 1954. URL http://link.aps.org/doi/10.1103/PhysRev.96.99.

[47] Tadao Kasuya. A Theory of Metallic Ferro- and Antiferromagnetism on Zener's Model. *Progress of Theoretical Physics*, 16(1):45–57, 1956. URL http://ptp.oxfordjournals.org/content/16/1/45.abstract.

[48] Kei Yosida. Magnetic Properties of Cu-Mn Alloys. *Phys. Rev.*, 106:893–898, Jun 1957. URL http://link.aps.org/doi/10.1103/PhysRev.106.893.

[49] Lihui Zhou, Jens Wiebe, Samir Lounis, Elena Vedmedenko, Focko Meier, Stefan Bl"ugel, Peter H. Dederichs, and Roland Wiesendanger. Strength and directionality of surface Ruderman-Kittel-Kasuya-Yosida interaction mapped on the atomic scale. *Nature Physics*, 6(3):187–191, 2010. URL http://dx.doi.org/10.1038/nphys1514.

[50] *Gmelin-Handbuch der anorganischen Chemie*, volume Seltene Erden ; B4: Seltenerdelemente: B4Sc, Y, La und Lanthanide : Eigenschaften der Kerne, Atome, Moleküle. Springer, Berlin, 8th edition, 1976.

[51] Marina Pivetta, Markus Ternes, Fran çois Patthey, and Wolf-Dieter Schneider. Diatomic Molecular Switches to Enable the Observation of Very-Low-Energy Vibrations. *Phys. Rev. Lett.*, 99(12):126104, 2007. URL http://link.aps.org/doi/10.1103/PhysRevLett.99.126104.

[52] F. Donati, A. Singha, S. Stepanow, C. Wäckerlin, J. Dreiser, P. Gambardella, S. Rusponi, and H. Brune. Magnetism of ho and er atoms on close-packed metal surfaces. *Phys. Rev. Lett.*, 113:237201, 2014. URL http://link.aps.org/doi/10.1103/PhysRevLett.113.237201.

[53] Cyrus F. Hirjibehedin, Christopher P. Lutz, and Andreas J. Heinrich. Spin Coupling in Engineered Atomic Structures. *Science*, 312:1021–1024, 2006. URL http://dx.doi.org/10.1126/science.1125398.

[54] A. F. Otte, M. Ternes, S. Loth, C. P. Lutz, C. F. Hirjibehedin, and A. J. Heinrich. Spin Excitations of a Kondo-Screened Atom Coupled to a Second Magnetic Atom. *Phys. Rev. Lett.*, 103(10):107203, 2009. URL http://dx.doi.org/10.1103/PhysRevLett.103.107203.

[55] F. M. Leibsle, C. F. J. Flipse, and A. W. Robinson. Structure of the Cu100-c(2×2)N Surface: A Scanning-Tunneling-Microscopy Study. *Phys. Rev. B*, 47:15865–15868, 1993.

[56] F.M. Leibsle, S.S. Dhesi, S.D. Barrett, and A.W. Robinson. STM Observations of Cu(100)-c(2×2)N Surfaces: Evidence for Attractive Interactions and an Incommensurate c(2×2) Structure. *Surf. Sci.*, 317:309–320, 1994. URL http://dx.doi.org/10.1103/PhysRevB.47.15865.

[57] Markus Ternes, Andreas J. Heinrich, and Wolf-Dieter Schneider. Spectroscopic manifestations of the Kondo effect on single adatoms. *Journal of Physics: Condensed Matter*, 21(5):053001, 2009. URL http://stacks.iop.org/0953-8984/21/i=5/a=053001.

[58] Werner A Hofer, Gilberto Teobaldi, and Nicolas Lorente. Creating pseudo-Kondo resonances by field-induced diffusion of atomic hydrogen. *Nanotechnology*, 19 (30):305701, 2008. URL http://stacks.iop.org/0957-4484/19/i=30/a=305701.

[59] Fabien Silly, Marina Pivetta, Markus Ternes, Fran çois Patthey, Jonathan P. Pelz, and Wolf-Dieter Schneider. Creation of an Atomic Superlattice by Immersing Metallic Adatoms in a Two-Dimensional Electron Sea. *Phys. Rev. Lett.*, 92: 016101, Jan 2004. URL http://link.aps.org/doi/10.1103/PhysRevLett.92.016101.

[60] Arnold F. Holleman, Egon Wiberg, and Nils [Editor] Wiberg. *Lehrbuch der anorganischen Chemie*. de Gruyter, Berlin, 102nd edition, 2007.

[61] A. Spinelli, B. Bryant, F. Delgado, J. Fernndez-Rossier, and A. F. Otte. Imaging of spin waves in atomically designed nanomagnets. *Nat. Mater.*, 13(8):782–785, 2014. URL http://dx.doi.org/10.1038/nmat4018.

[62] S. M. York and F. M. Leibsle. Co nanowire arrays on *N* -terminated Cu(110) surfaces. *Phys. Rev. B*, 64:033411, Jun 2001. URL http://link.aps.org/doi/10.1103/PhysRevB.64.033411.

[63] X.-D. Ma, D. I. Bazhanov, O. Fruchart, F. Yildiz, T. Yokoyama, M. Przybylski, V. S. Stepanyuk, W. Hergert, and J. Kirschner. Strain Relief Guided Growth of Atomic Nanowires in a Cu_3N-Cu(110) Molecular Network. *Phys. Rev. Lett.*, 102:205503, 2009. URL http://link.aps.org/doi/10.1103/PhysRevLett.102.205503.

[64] Sebastian Wedekind, Fabio Donati, Hirofumi Oka, Guillemin Rodary, Dirk Sander, and Jürgen Kirschner. The electronic properties of co nanowires on cu(110)-p(2×3)n. *Surface Science*, 606(2122):1577 – 1580, 2012. URL http://www.sciencedirect.com/science/article/pii/S0039602812002051.

[65] F. M. Leibsle, R. Davis, and A. W. Robinson. Scanning-tunneling-microscopy studies of (2×3)N-induced structures, thermal desorption, and oxygen coadsorption on the Cu110 surface. *Phys. Rev. B*, 49:8290–8299, Mar 1994. URL http://link.aps.org/doi/10.1103/PhysRevB.49.8290.

[66] B. Bryant, R. Toskovic, A. Ferrn, J. L. Lado, A. Spinelli, J. Fernández-Rossier, and A. F. Otte. Controlled complete suppression of single-atom inelastic spin and orbital cotunnelling. 2015. URL http://arxiv.org/abs/1504.05882.

Experimental Condensed Matter Physics
(ISSN 2191-9925)

Die Bände sind unter www.ksp.kit.edu als PDF frei verfügbar
oder als Druckausgabe bestellbar.

Band 1 Alexey Feofanov
Experiments on flux qubits with pi-shifters. 2011
ISBN 978-3-86644-644-1

Band 2 Stefan Schmaus
Spintronics with individual metal-organic molecules. 2011
ISBN 978-3-86644-649-6

Band 3 Marc Müller
Elektrischer Leitwert von magnetostriktiven Dy-Nanokontakten. 2011
ISBN 978-3-86644-726-4

Band 4 Torben Peichl
Einfluss mechanischer Deformation auf atomare Tunnelsysteme – untersucht mit Josephson Phasen-Qubits. 2012
ISBN 978-3-86644-837-7

Band 5 Dominik Stöffler
Herstellung dünner metallischer Brücken durch Elektromigration und Charakterisierung mit Rastersondentechniken. 2012
ISBN 978-3-86644-843-8

Band 6 Tihomir Tomanic
Untersuchung des elektronischen Oberflächenzustands von Ag-Inseln auf supraleitendem Niob (110). 2012
ISBN 978-3-86644-898-8

Band 7 Lukas Gerhard
Magnetoelectric coupling at metal surfaces. 2013
ISBN 978-3-7315-0063-6

Experimental Condensed Matter Physics
(ISSN 2191-9925)

Band 8 Kirill Fedorov
Fluxon readout for superconducting flux qubits. 2013
ISBN 978-3-7315-0067-4

Band 9 Jochen Zimmer
Cooper pair transport in arrays of Josephson junctions. 2014
ISBN 978-3-7315-0130-5

Band 10 Oliver Berg
Elektrischer Transport durch Nanokontakte von Selten-Erd-Metallen. 2014
ISBN 978-3-7315-0209-8

Band 11 Grigorij Jur'evic Grabovskij
Investigation of coherent microscopic defects inside the tunneling barrier of a Josephson junction. 2014
ISBN 978-3-7315-0210-4

Band 12 Cornelius Thiele
STM Characterization of Phenylene-Ethynylene Oligomers on Au(111) and their Integration into Carbon Nanotube Nanogaps. 2014
ISBN 978-3-7315-0235-7

Band 13 Michael Peter Schackert
Scanning Tunneling Spectroscopy on Electron-Boson Interactions in Superconductors. 2014
ISBN 978-3-7315-0238-8

Band 14 Susanne Butz
One-Dimensional Tunable Josephson Metamaterials. 2014
ISBN 978-3-7315-0271-5

Band 15 Philipp Jung
Nonlinear Effects in Superconducting Quantum Interference Meta-Atoms. 2014
ISBN 978-3-7315-0294-4

Band 16 Sebastian Probst
Hybrid quantum system based on rare earth doped crystals. 2015
ISBN 978-3-7315-0345-3

Experimental Condensed Matter Physics
(ISSN 2191-9925)

Band 17 Wolfram Kittler
Magnetische Anisotropie und Quantenphasenübergang in $CeTi_{1-x}V_xGe_3$. 2015
ISBN 978-3-7315-0363-7

Band 18 Moritz Peter
Towards magnetic resonance in scanning tunneling microscopy using heterodyne detection. 2015
ISBN 978-3-7315-0410-8

Band 19 Junji Tobias Märkl
Investigation of Magnetic Adatoms with Scanning Tunneling Techniques. 2015
ISBN 978-3-7315-0435-1